Y0-BYK-274

# BIOCHEMISTRY OF CELL DIFFERENTIATION

# Seventh FEBS Meeting

**Volume 22**

VIRUS-CELL INTERACTIONS AND VIRAL ANTIMETABOLITES

**Volume 23**

FUNCTIONAL UNITS IN PROTEIN BIOSYNTHESIS

**Volume 24**

BIOCHEMISTRY OF CELL DIFFERENTIATION

FEDERATION OF EUROPEAN BIOCHEMICAL SOCIETIES
SEVENTH MEETING, VARNA (BULGARIA), SEPTEMBER 1971

# BIOCHEMISTRY OF CELL DIFFERENTIATION

Volume 24


*Edited by*

A. MONROY

*CNR Laboratory of Molecular Embryology,*
*Naples, Italy*

AND

R. TSANEV

*Biochemical Research Laboratory,*
*Bulgarian Academy of Sciences,*
*Sofia, Bulgaria*

1973

ACADEMIC PRESS . London and New York

ACADEMIC PRESS INC. (LONDON) LTD.
24/28 Oval Road,
London NW1

*United States Edition published by*
ACADEMIC PRESS INC.
111 Fifth Avenue
New York, New York 10003



Library of Congress Catalog Card Number: 72-12275
ISBN: 0-12-504 850-5

Printed in Great Britain by
William Clowes & Sons Limited
London, Colchester and Beccles

## Introductory Remarks

The Biochemistry of Cell Differentiation is a vast and indeed very rapidly-evolving field. In recent years, a constant flow of new observations and experimental results has impelled continual readjustment of our thinking. Hence, a major difficulty in organizing a Symposium lies in the possibility that topics selected for eventual discussion may already be obsolete by the time the meeting actually takes place. We were fortunate enough, in a sense, that no such phenomenon occurred to disturb our Symposium; indeed, the questions discussed are more than ever before in the forefront of the study of cell differentiation.

When the problem of cell differentiation is approached at the biochemical level, the first requirement is a biochemical marker of the differentiated state. And what one specifically wishes to have are protein markers. In fact, at the molecular level, differentiation is assumed to imply the synthesis of specific proteins. Several such markers have been with us for many years, such as haemoglobin, the muscle proteins or some of the glandular secretion products, but only very recently has it become possible to use them as tools for the study of differentiation at the molecular level. In recent years, other proteins and enzymes have also proved to be extremely valuable markers. One of these is the enzyme glutamine synthetase as a marker of the differentiation of the neural retina. This enzyme, in the hands of our first speaker, A. A. Moscona, has yielded some very important information concerning both transcriptional and translational controls. One of the most interesting aspects of this enzyme is the possibility of turning it on in advance with respect to its normal temporal programme by the use of certain hormones.

The erythropoietic and myogenic systems have been used by H. O. Holtzer and his colleagues with the aim of obtaining an answer to one of the central problems of differentiation, namely that of the relationships between DNA replication and the expression of a certain differentiation programme. Specifically, the question is the following: what is the nature of the switch that causes certain cells at one point either to stop dividing (as in the case of the myogenic cells) or, while still dividing (as in the case of the haemocytoblasts) to undertake synthesis of certain specific proteins? The switch must obviously mean that a new genetic programme now channels the synthetic machinery of the cell into a new pathway.

The erythropoietic system lends itself also to pushing the analysis one step farther. The isolation of the haemoglobin messenger-RNA now allows a study of the chain of events leading from the transcription of the messenger precursors through their processing to their eventual translation. Work in recent years has shown that in eukaryotes the gene products need to undergo an intra-nuclear processing similar, as is now well known, to what occurs in the case of the ribosomal RNA, and which appears to be a prerequisite for their transport to and utilization in the cytoplasm. How much of the high molecular weight RNA, which is the primary gene product, does in fact leave the nucleus? And how is the selection operated between the segments that are to be transported and those destined instead to be degraded within the nucleus? To these questions, both K. Scherrer's group and G. P. Georgiev's are seeking an answer.

One of the most striking discoveries in the last few years has been that of gene amplification, first demonstrated with certainty in amphibian oocytes, where amplification may be of the order of several thousand times. This degree of amplification is such as to make this material an ideal one to ask a number of questions related not only to gene amplification but also to the regulation of gene activity in general. And it is particularly with this in mind that the opportunities offered by this system are being exploited by M. Crippa and G. P. Tocchini-Valentini. The discovery of gene amplification has also prompted the question as to whether or not gene amplification plays a more general role and, more specifically, whether it actually plays a role in differentiation. One obvious system of which to ask this question is again the erythropoietic system, and especially the reticulocytes of anaemic animals, in which essentially only one kind of protein, haemoglobin, is synthesized. This problem is being investigated by J. Bishop and his colleagues, who are availing themselves of the high resolving power of the DNA/RNA hybridization technique they have worked out.

A classical object for the study of the control of gene activity is still the giant chromosomes of the Diptera. In recent years, a number of technical refinements have brought this material within the realm of analysis at the molecular level; and Dr. C. Pelling will bring us up-to-date on this subject.

As I mentioned at the beginning of this introduction, glands, and the pancreas in particular, offer interesting possibilities for the study of differentiation control at the biochemical level. Indeed, starting from the early rudiment, full differentiation of the pancreatic structure can be obtained *in vitro*: and what is more important, the technique of *in vitro* culture, as exploited by W. Rutter and his colleagues, has allowed them to investigate the role of interaction with the mesenchyme in the differentiation control of the organ. Since embryonic induction is the result of tissue interaction, this work is most pertinent to one of the major and still unsolved problems of developmental biology.

## INTRODUCTORY REMARKS

What I have tried to do in this Introduction is not to present you with an outline of the selected abstracts, but rather to indicate to you the general trend Professor R. Tsanev and I have followed in organizing this Symposium. Indeed, one of the difficulties of a Symposium like this is to be constrained to select only a few out of a great number of equally important topics. Understandably, the choice is inevitably influenced by the personal inclinations of the organizers. On behalf also of Professor Tsanev, I wish to thank all the participants for having accepted our invitation; we feel we have been very lucky indeed to have succeeded in bringing together some of the most outstanding leaders in the field.

*Varna, Bulgaria* A. MONROY

NOTE: The Editors would like to express their thanks to the publishers for facilitating the publication of this volume, notwithstanding the difficulties arising out of the delay in collecting and furnishing the manuscripts.

# Contents

FEBS Symposium, Volume 24, 1972, pp. 1-23

# Induction of Glutamine Synthetase in Embryonic Neural Retina: A Model for the Regulation of Specific Gene Expression in Embryonic Cells

A. A. MOSCONA

*Department of Biology, The University of Chicago, Chicago, Illinois, U.S.A.*

## INTRODUCTION

For the purpose of this discussion embryonic differentiation is considered as cellular changes in macromolecular synthesis and composition, patterned in time and space and resulting in specialized cell functions, forms and organizations. There is ample evidence that these changes cannot be ascribed to gross differences in the genetic make-up of the cells and that they predominantly represent differences in gene expression in genetically equipotential genomes. Therefore, mechanisms must exist which specify which genes are to be expressed in given cells at particular times in development; identification of these mechanisms and their modes of action is a major objective of studies on differentiation.

It is known that signals from outside the cell play an important role in the regulation of growth and differentiation. Such signals may be transmitted through contact between cells, by intercellular junctions, or by cell membrane interactions; or, they may be transferred across greater distances by molecules which originate in cells of one kind and affect other cells. The target cell genome reacts to the signals from the outside and in so doing modifies the surrounding environment, and it is such reciprocal, continuous interactions between genome, cytoplasm, and extracellular milieu that propels the cell along the course of differentiation.

The process by which external factors elicit changes in embryonic cells resulting in tissue-specific synthesis is referred to as *embryonic induction*. To be suitable for analysis of induction, an embryonic system should meet the following experimental requirements: it should consist of a well characterized population of embryonic cells in which a defined effector can induce, under controlled conditions, the formation of a product characteristic for the differentiation of these cells; ideally, it would be desirable to elicit such an induction precociously, at an earlier phase of development than that at which it normally occurs, since specific modifications of timetables of developmental processes are essential to the unravelling of mechanisms of differentiation.

At present, very few of the factors which trigger the synthesis of tissue-specific proteins during development have been conclusively identified, but a number of experimental systems show promise in this direction; among these are the induction of hemoglobin by erythropoietin in hemopoietic cells (Goldwasser, 1966; Stephenson *et al.*, 1971); formation of lens proteins in amphibian lens regeneration from the iris (Yamada, 1967); synthesis of cell-wall enzymes in slime molds (Sussman, 1966); hormonal regulation of insect and amphibian metamorphosis (Tata, 1971); regulation of ovalbumin and avidin production in the oviduct (Dingman *et al.*, 1969) and of casein in the mammary gland (Turkington, 1968); induction of pancreatic enzymes (Rutter and Weber, 1965), etc.

Even less common are embryonic systems in which it is possible to induce precociously specific changes in patterns of protein synthesis, well before the time of their regular appearance in the course of normal differentiation. One of the few is the induction of glutamine synthetase in the embryonic chick neural retina. This system meets the essential requirements for analysis of induction, mentioned above; glutamine synthetase is a marker of differentiation in the normal retina and its regulation in this tissue by a chemically known effector provides a model for studying control mechanisms of general significance to development.

## THE DEVELOPMENTAL PATTERN OF GLUTAMINE SYNTHETASE IN THE EMBRYONIC CHICK RETINA

Glutamine synthetase [L-glutamate: ammonia ligase (ADP)], to be referred to as *GS*, catalyses the conversion of glutamate to glutamine and is, therefore, important in various biosynthetic pathways including those for nucleotides and polysaccharides. In nerve cells GS is of special interest in that it is associated with membranes and synaptic structures (Salganicoff and deRobertis, 1965) and because of the postulated role of glutamate-glutamine in neural transmission. Mammalian brain GS, first purified in Meister's laboratory (Meister, 1968; Ronzio *et al.*, 1969) was described to have a molecular weight of approximately 392,000 and to consist of eight apparently identical subunits (Tate and Meister, 1971). GS purified from chicken retina has an estimated subunit molecular weight of approximately 42,000 (Sarkar *et al.*, 1972).

The neural retina develops in the embryo from a relatively simple epithelial structure into a complex tissue specialized for the reception and transmission of light signals. Among the biochemical changes in the differentiation of the retina that are uniquely characteristic for its developmental program is a sharp increase of GS activity (Fig. 1).* During early development of the chick embryo, GS activity in the neural retina is low and increases very slowly; on about the 16th day of embryonic life it begins to rise very rapidly and increases approximately a hundredfold in 5 days, then plateaus and remains at this high level probably throughout adult life (Moscona and Hubby, 1963; Moscona,

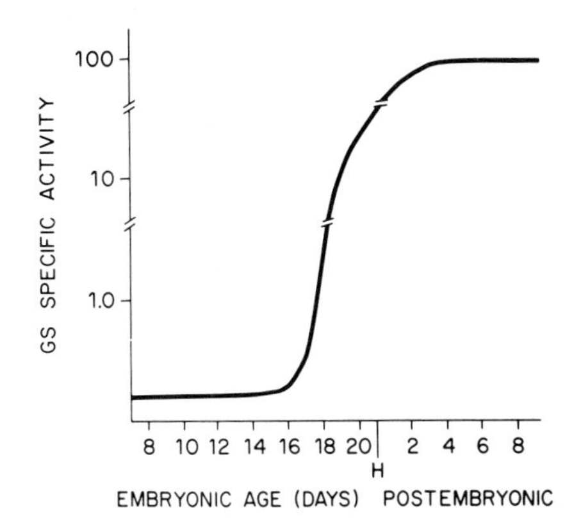


**Figure 1.** The developmental pattern of glutamine synthetase (GS) specific activity in the neural retina of the chick embryo and after hatching (H).

1971). This marked increase of GS activity coincides with the period of maturation in the neural retina; in fact, there is an excellent correlation between the increase of GS activity in the neural retina and its functional differentiation, not only in the chick embryo (Piddington and Moscona, 1965) but also in mammalian eye development (Chader, 1971). In no other tissue of the chick embryo does GS activity show a similar pattern of developmental change; only in certain parts of the brain associated with vision does it increase significantly at about the same time as in the retina (Shimada *et al.*, 1967; Piddington, 1971).

The rapid increase of GS activity on the 16th day is not due to cell proliferation, and is not associated with marked changes in the total protein content of the retina. Figure 2 shows that growth of the neural retina practically ceases after the 12th day of development: both cell number and total protein content increase rapidly before the 12th day, but level off thereafter. Therefore, on the 16th day the retina is essentially a non-growing tissue, engaged predominantly in differentiation.

* The activity of GS is assayed routinely in tissue homogenates or sonicates by the glutamyl transferase reaction (see: Kirk and Moscona, 1963; Moscona *et al.*, 1968).

The developmental pattern of GS activity in the chick embryo retina raises general questions which apply also to other differentiating systems. First, is the same enzyme responsible for the base level and the high level activities, or are they due to different enzyme proteins. Second, is the rapid increase of GS activity caused by an intrinsic mechanism within the retina cells, or by an external factor of systemic or hormonal nature. Third, can GS be caused to increase rapidly before the normal time, or must the retina reach the age of 16 days in order to show this developmental change. And fourth, is the increase of the enzyme activity due to enzyme synthesis, or to activation of performed precursors, and how is this increase controlled by the genome.

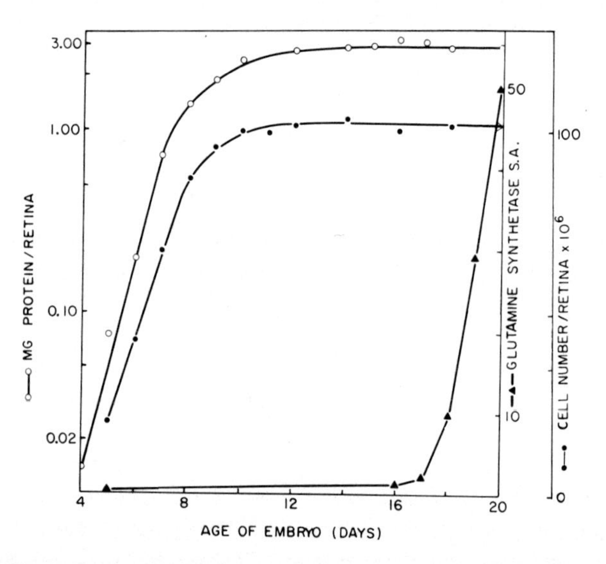


**Figure 2.** Glutamine synthetase (GS) activity in the neural retina of the chick embryo increases sharply after the 16th day of embryonic development. Cell number and total protein content in the neural retina increase till the 11-12th day of embryogenesis and do not change conspicuously thereafter.

I will anticipate my discussion by, first, answering the above questions briefly. Concerning the base level and the high level GS activities, by no test yet performed has it been possible to differentiate qualitatively between the properties of the enzyme at these levels; studies with antiserum against purified GS revealed no immunological differences between them; their identical electrophoretic mobility, sensitivity to irreversible inhibition by L-methionine-S-sulfoximine and other similarities (Kirk and Moscona, 1963) indicate that the base level and the high level GS activities are due to the same enzyme; this statement applies also to the precociously induced GS, as discussed further on. Second, a systemic factor induces the sharp increase of

GS activity in the retina and can elicit this effect in retinas much younger than 16 days, resulting in a premature appearance of a change programmed to start at a later stage in normal development. Third, the induction of GS activity in the retina is due to enzyme synthesis and to accumulation of newly made enzyme. Finally, the induction of GS requires gene function and synthesis of the enzyme is continuously controlled through gene action.

## PRECOCIOUS INDUCTION OF GS IN THE EMBRYONIC RETINA BY A SERUM FACTOR

The rapid increase of GS activity in the embryonic chick neural retina is triggered by a factor present in adult serum and in serum from late chick embryos. In early embryos this factor is either absent, or is below active level, and it becomes available shortly before the 16th day of development, i.e. before the onset of the sharp rise of GS activity in the retina. It is present in adult mammalian serum, but is low in fetal serum (unless it is late fetal, or is contaminated with adult serum). There is compelling evidence that this factor is an adrenal corticosteroid hormone (Piddington, 1967, 1970; Moscona, 1971).

That a factor present in adult serum can induce GS precociously in the retina was first demonstrated in experiments (Moscona and Hubby, 1963; Moscona and Kirk, 1965) using isolated embryonic chick retina tissue maintained in flask cultures in a defined culture medium.† Retinas from 10 through 16-day chick embryos were cultured for 24 h in buffered physiological salt solution (Tyrode) with adult serum, and were then assayed for GS activity. Even in the youngest of these retinas GS activity increased significantly in 24 h, showing that the embryonic retina possessed the ability to undergo a marked increase of GS activity long before the 16th day of development; without adult serum, or when fetal serum was used instead, there was no such increase. Adult sera from various organisms induced retinal GS precociously, so that the responsible factor was not of species-specific nature. On the other hand, the inductive effect of adult serum was rather tissue-specific in that it induced GS precociously, to this extent, only in the embryonic neural retina [and to a lesser extent in brain regions of the chick embryo associated with vision (Shimada *et al.*, 1967; Piddington, 1971)], but in no other tissue of the chick embryo yet tested.

† To obtain retina tissue, the eyes are cut into halves, the vitreous humor is removed, and the retina comes out cleanly as an avascular sheet of pure, neural cells. Retina from one eye of a 12-day, or older embryo contains approximately 3 mg protein and $10^8$ cells. The isolated retina tissue is placed in an Erlenmeyer flask with culture medium and is incubated (38°C) on a gyratory shaker rotating at 70 rpm (Moscona *et al.*, 1968).

## THE INDUCTION OF GS IN THE EMBRYONIC RETINA BY 11β-HYDROXYCORTICOSTEROIDS

Analysis of adult serum indicated that the factor inducing retinal GS was a corticosteroid. Screening of various corticosteroids for ability to induce GS in isolated retina from 12-day embryos, and examination of their dose response curves (Fig. 3) showed that certain 11β-hydroxycorticosteriods were excellent inducers of GS, when applied at low concentrations (Moscona and Piddington, 1966). These included corticosterone, aldosterone, hydrocortisone,

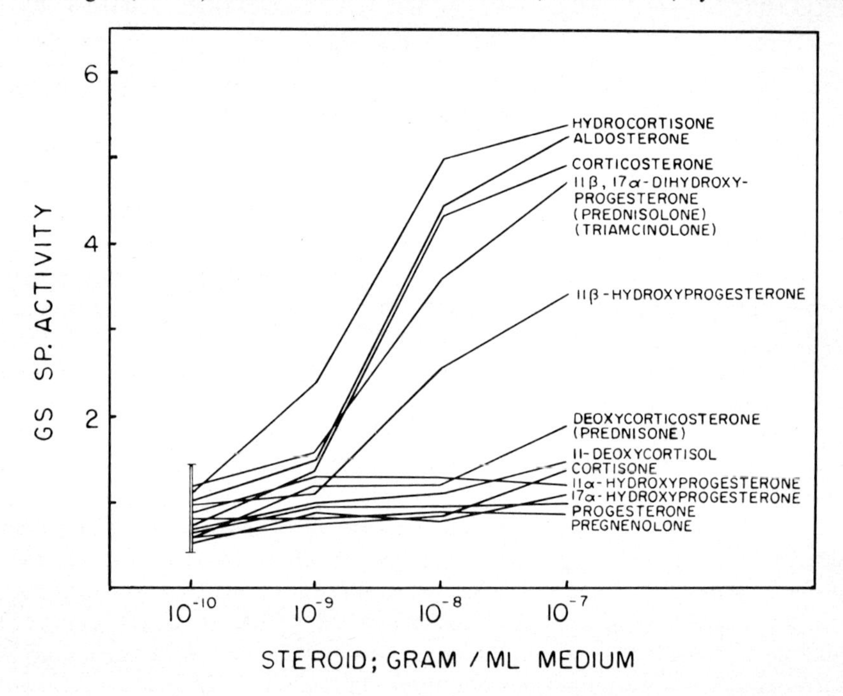


**Figure 3.** Precocious hormonal induction of GS in isolated neural retina from 12-day chick embryos: dose-response effects of various steroids during 24 hours in culture.

and structurally related synthetic steroids such as dexamethasone, prednisolone, and triamcinolone (Moscona, 1971). The inductive activity of these molecules was critically dependent on the presence of the 11-hydroxyl in β configuration, and was amplified by hydroxyls in position 21 (Moscona and Piddington, 1967). In addition, the hydroxyl in position 3 was also essential for the inductive effect, possibly because of its postulated role in the transport of the steroid into the cell (Reif-Lehrer, 1968; Reif-Lehrer and Amos, 1968).

Figure 4 shows the precocious induction of GS by hydrocortisone in 24-h cultures of retina tissue from 7 through 16 day-old chick embryos; the steroid induced a several-fold increase in GS activity in all cases, while in controls without the steroid there was no induction. The culture medium contained only physiological salt solution (Tyrode) and the steroid, showing that the steroid was the actual inducer of GS and that GS induction in the retina under these conditions did not depend on provision of other exogenous micro- or macromolecular requirements (Moscona *et al.*, 1968; Moscona, 1971).

The precocious induction of GS in the retina is not limited to culture conditions; the active steroids can induce retinal GS also *in vivo*, when

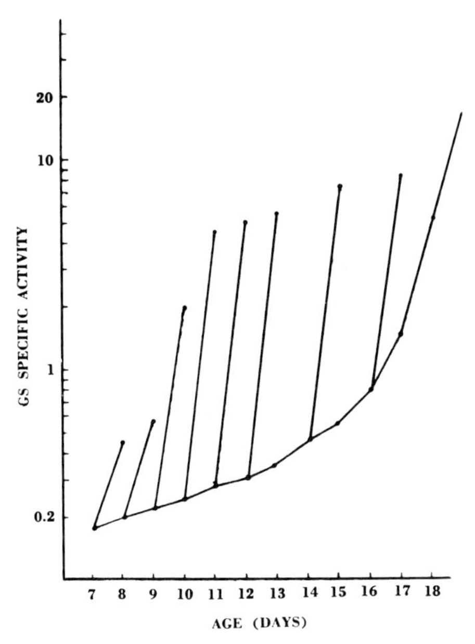


**Figure 4.** Precocious induction of GS by hydrocortisone ($0.3 \times 10^{-7}$ g/ml) in neural retina isolated from chick embryos 7 to 16 days old and maintained *in vitro* for 24 h in the presence of the steroid inducer. Baseline: GS specific activities in the developing retina *in situ.* Vertical lines: Increases in GS specific activity induced by the steroid in isolated retina of different embryonic eyes during 24 h *in vitro.* (From Piddington and Moscona, 1967.)

injected into early embryos as early as 9 days before the normal rise of this enzyme (Piddington and Moscona, 1967). The fact that supplying the embryo prematurely with the appropriate steroid elevates GS in the retina precociously *in situ* further supports the suggestion that an endogenous corticosteroid is normally responsible for the increase of GS on the 16th day of embryonic development. There is evidence that in the chick embryo the function of the adrenal cortex increases appreciably after the 14th day of development (Piddington, 1970). Furthermore, injections of adrenocorticotrophic hormone (ACTH) into young chick embryos to stimulate premature function of the adrenal cortex elicited precocious increases of GS in the retina *in situ* (but not when ACTH was applied to retina tissue in culture); on the other hand, inhibition of steroid biogenesis in the adrenal cortex reduced or retarded its ability to promote GS increases in the retina (Piddington, 1970).

Other hormones tested so far failed to induce retinal GS precociously, either in culture or in the embryo; these include insulin, glucagon, adrenalin and sex hormones. Cyclic AMP and dibutyryl cyclic AMP also did not elevate GS in the embryonic retina above minor increases of the baseline level, nor did they amplify the effect of the inductive steroids (Moscona, 1971).

Therefore, the present evidence, while short of direct proof, points compellingly to the conclusion that a corticosteroid in the embryo is the natural inducer of GS in the embryonic retina. The present working assumption is that, in early embryos the level of this steroid is low, and that its concentration, or availability increases in later development and upon reaching an effective threshold GS in the retina is induced. This does not exclude the possible existence also of intracellular autonomous regulatory mechanisms of GS, or of other serum factors which affect the inducibility of GS in the retina. (For further aspects of this problem see Moscona and Kirk, 1965; Moscona, 1971.)

The fact that the retina cells are responsive to GS induction by the steroid inducer 8-9 days before the enzyme rises in the embryo, implies not only that cellular receptors for the steroid and mechanisms for its intracellular transport exist in these cells long before the time they are normally used in GS induction, but also that the genes involved in the regulation of GS can react to this specific hormonal signal days before they are normally due for this action and while the cells are still in an extremely premature morphological, numerical and physiological state. This raises the interesting issue of "competence" to induction in various embryonic systems, in which previous attempts have failed to elicit precociously, significant developmental changes. In view of the precocious induction of GS, it seems debatable whether such negative results are due to a true non-responsiveness of the cells and indicate non-readiness of the genome, as often suggested, or, whether they reflect primarily unawareness of the specific signals that cause the specific changes.

## MACROMOLECULAR SYNTHESIS IN GS INDUCTION

### Protein synthesis

Turning now to the mechanisms of GS induction, Fig. 5 shows the induction of GS in cultured retina from 12-day embryos by $10^{-7}$ g/ml hydrocortisone; after a short lag, GS activity increases in 24 h to a value approximately 10 times higher than in controls without the inducer. The induction of GS requires continuous protein synthesis and is prevented if cycloheximide or puromycin at concentrations which stop protein synthesis in these cells are added together with the inducer; addition of these inhibitors at any time after the beginning of induction stops, reversibly, further increases of enzyme activity and causes it to level off (Moscona *et al.*, 1968).

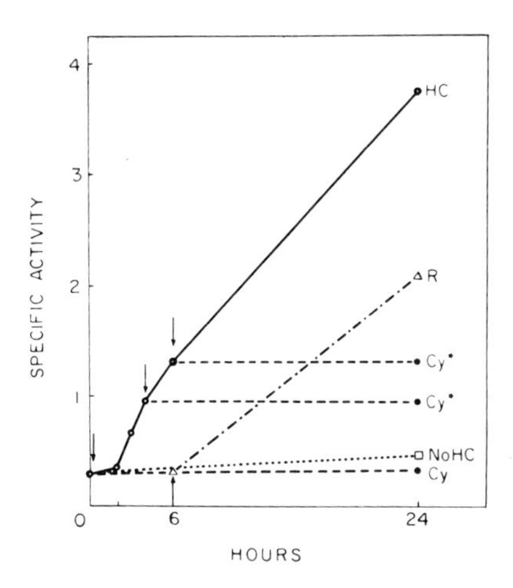


**Figure 5.** The effect of cycloheximide (Cy) on GS induction by hydrocortisone (HC) in neural retina from 12-day chick embryo. HC was added at zero hour to all cultures, except the one marked "No HC." Arrows pointing down mark additions of Cy (2 μg/ml). Arrow pointing up marks withdrawal time of Cy to show the reversibility (R) of its effect. Addition of the inhibitor after 4 or 6 h of induction caused the GS activity to level off (Cy*).

Although these data suggest that GS induction might be due to enzyme synthesis, they do not exclude other possibilities. For example, the induction by cortisol of alkaline phosphatase in HeLa cells is also inhibited by cycloheximide, however the blocked protein is not the actual enzyme, but probably a molecule which enhances the catalytic activity of preformed enzyme (Cox *et al.*, 1971); the cortisol-mediated increase of alkaline phosphatase activity in embryonic mouse intestine also is due to activation of preformed enzyme (Etzler and Moog, 1966).

In the case of GS induction, no evidence was found that it was due to activation of preformed enzyme, or to a change in the concentration of a competitive inhibitor (Kirk and Moscona, 1963). Direct evidence that GS induction involves enzyme synthesis was obtained by immunotitration of the enzyme, and by analysis at the polysomal level.

Using specific antiserum against purified GS, immunoprecipitation measurements were carried out, comparing synthesis and accumulation of radioactively

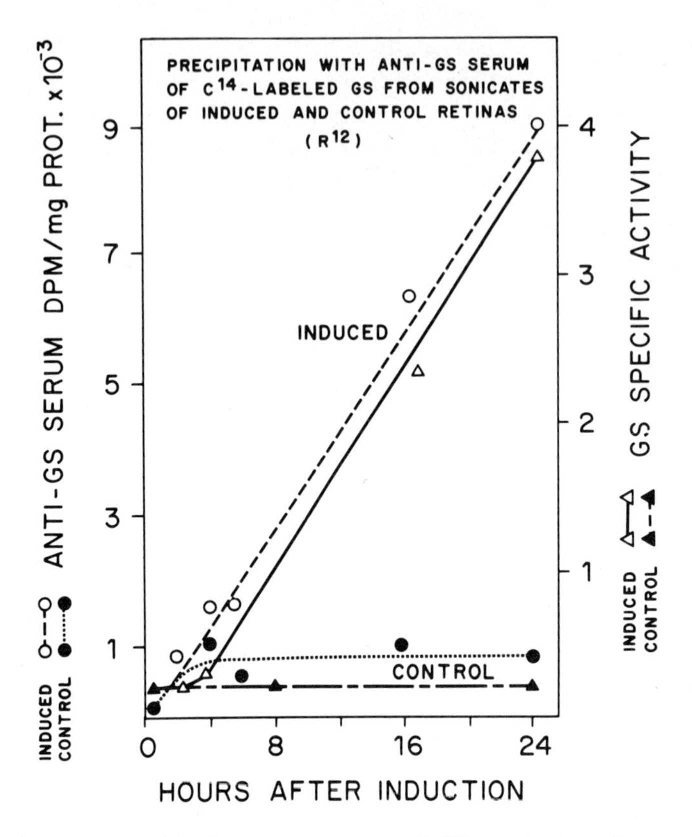


**Figure 6.** Immunoprecipitation measurements of GS synthesis and accumulation induced by HC in embryonic neural retina. Further details in text. (Modified from Alescio and Moscona, 1969.)

labeled enzyme in induced and non-induced retina.§ These measurements showed (Fig. 6) that induction of GS represented synthesis and accumulation of the enzyme, and that there was a close correspondence between increases in amount of immunoprecipitable labeled GS and increases of enzyme activity (Alescio and Moscona, 1969; Moscona *et al.*, 1972). Since in some tissues corticosteroids affect overall protein synthesis (Kenney, 1970; Manchester, 1970) it should be pointed out that in the retina there is no measurable

§ For immunoprecipitation tests, retina was cultured with and without the inducer, in the presence of $^{14}C$ amino acids; after various culture times, the tissue was sonicated and the 100,000 g supernatant was reacted with antiserum against GS purified from sheep brain or from chicken retina; the amount of radioactivity precipitated by the antiserum provided measures of enzyme synthesis and accumulation (Alescio and Moscona, 1969; Moscona *et al.*, 1972).

difference in overall protein synthesis or content between induced and non-induced tissue; thus, the induction of GS in the retina represents a differential biosynthetic response of these cells to this inducer (Alescio and Moscona, 1969).

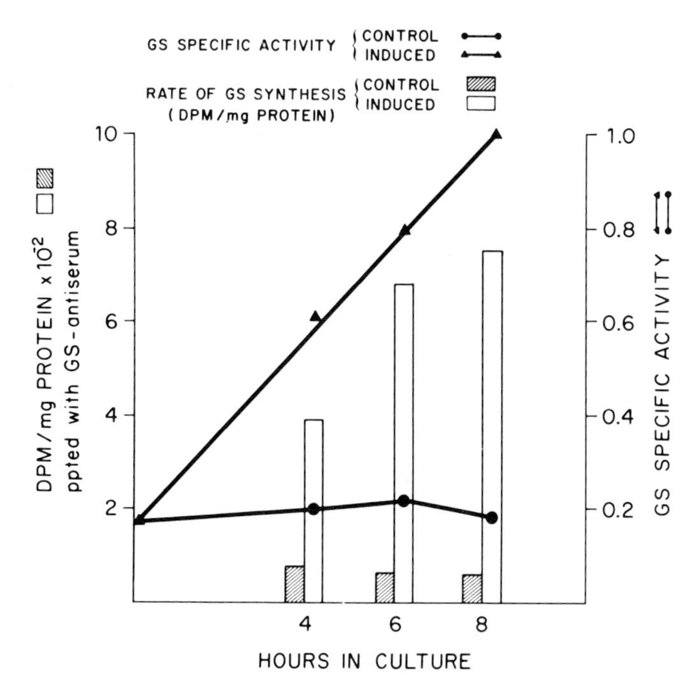


**Figure 7.** Induction of GS in embryonic neural retina: changes in the rate of enzyme synthesis measured by immunotitration of the radioactively labeled enzyme (for explanation see text). Comparisons of GS specific activities and rates of GS synthesis, in induced and control retinas. For further details see text. (From Moscona *et al.*, 1972.)

A further important fact brought out by immunoprecipitation measurements was that GS is continuously made also in the non-induced retina; however, the enzyme does not accumulate significantly, undoubtedly because it is continuously degraded at a rate which balances its rate of synthesis. This raises the fundamental question whether the rapid accumulation of GS in induction is due to a marked increase in the rate of enzyme synthesis or, predominantly to inhibition of its turnover. This question is basic to the mechanism of induction; applies to induced accumulation of specific products in other embryonic systems, but very little information exists on this problem.

In the case of GS, immunoprecipitation measurements have shown that induction involves a progressive increase in the rate of enzyme synthesis and that enzyme degradation does not cease (Moscona *et al.*, 1972). In these experiments retinas were cultured with and without the inducer and after different culture times, were pulsed for 15 minutes with $^{14}$C-amino acids. The amount of $^{14}$C-labeled enzyme made during each of these short pulse periods was determined by immunoprecipitation to obtain rates of enzyme synthesis at different times of induction. The results (Fig. 7) showed that in the presence of the inducer the rate of GS synthesis increased with time; after 8 hours of induction it was 12 to 15-fold greater than in the non-induced

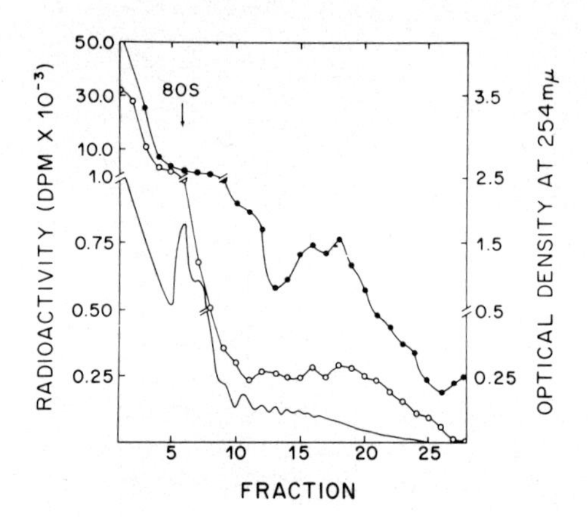


**Figure 8.** Sucrose-gradient (15-40%) analysis of 15,000 x g supernatants from control and hydrocortisone-induced retinas labeled in culture for 15 min with $^{14}$C Asp (3.3 μCi/ml), (see text). Absorbance profiles (solid line) were identical in both cases. o——o, Radioactivity (control); •——• Radioactivity (induced).

controls. Furthermore, enzyme turnover does not cease during induction; it undoubtedly plays a significant role in the balance of processes which determine the absolute amounts of GS available in cells during and after differentiation. The regulation of GS turnover requires further detailed study, especially in view of the evidence that it depends on continuous transcription (see below).

Analysis of retina polysomes for the presence of nascent GS provided further evidence concerning enzyme synthesis in GS induction (Sarkar and Moscona, 1971). Retinas were cultured in the presence or absence of the inducer, and were then pulsed for 15 minutes with $^{14}$C-aspartic acid. This amino acid was used because it is abundant in GS and its incorporation in induced retina is greater than in non-induced tissue. The polysomes were then isolated, fractionated on sucrose density gradients and the radioactivity in peptides was determined (Fig. 8). There were no differences between the $OD_{260}$ profiles, but in the induced retina profile there was a significantly greater incorporation of the label and a zone of elevated radioactivity in the region of 12-15 ribosome polymers which corresponds to peptide chains with a molecular weight of 35,000-45,000; this coincides with the estimated molecular weight of GS subunit which is 42,000 (Sarkar *et al.*, 1972). Therefore, assuming a monocistronic mRNA for GS, it is likely that the radioactivity in the region of 12-15 ribosome polymers represents nascent GS peptides. This conclusion is supported by various evidence that such increased incorporation of aspartate is closely correlated with GS induction, and that it is not elicited by the steroid in non-inducible tissues (Sarkar and Moscona, 1971).

**Nucleic acid synthesis**

*DNA synthesis.* Inhibition of ongoing DNA simultaneously with the addition of the inducer does not prevent GS induction (Moscona *et al.*, 1970). Furthermore, prolonged inhibition of DNA synthesis before the addition of

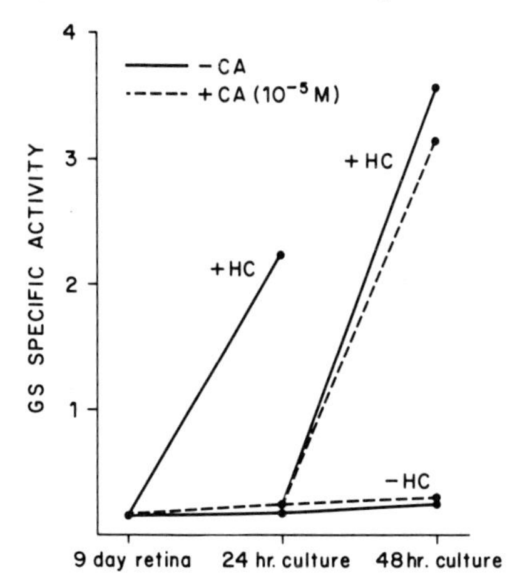


**Figure 9.** Effect of continuous inhibition of DNA synthesis in embryonic neural retina on inducibility of GS. Retinas isolated from 9-day chick embryos were cultured in the presence of $10^{-5}$ M cytosine arabinoside (CA) and without it; to some of the cultures, hydrocortisone was added at zero hour ($0.3 \times 10^{-7}$ g/ml); to others it was added after 24 hours. Controls received no steroid. The data show that continuous inhibition for 48 h of DNA synthesis did measurably reduce the responsiveness of the retina to GS induction.

the inducer did not measurably reduce the inducibility of GS in the retina. Cytosine arabinoside at $10^{-5}$ M stops rapidly and completely the incorporation of thymidine into DNA in retinas of all embryonic ages. In the experiments in Fig. 9, retinas from 9-day embryos were treated in culture for 24 h with cytosine arabinoside and were then exposed to the inducer for the next 24 hours in the continuous presence of the inhibitor. This persistent inhibition of DNA synthesis for a 48-h period, during which cell number in the normal retina would have increased considerably, did not measurably reduce the inducibility of the tissue and, following addition of the inducer GS activities rose to levels similar to those induced in non-inhibited retina. Therefore, even the 9-day retina can respond to the inductive stimulus without further total or partial replication of the genome. This situation differs from that described for some other systems, such as hormonal regulation of casein synthesis in the postnatal mammary gland, where cell replication is essential for acquisition of responsiveness to the hormonal effect (Mills and Topper, 1970). It also differs from the induction of tyrosine transaminase in hepatoma cell cultures, in which cell replication is prerequisite for inducibility (Tomkins, 1969).

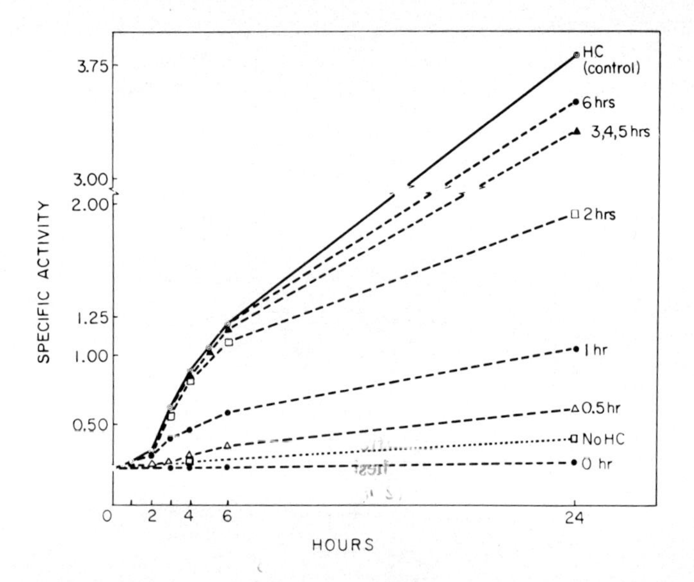


**Figure 10.** Effects of Act D (10 $\mu$g/ml) added at various times after the beginning of induction, on the induced increase of GS specific activity in 12-day embryonic retina. All cultures (except control-No HC) were with the inducer (HC). Act D was added at times listed on the right, next to the points indicating GS specific activity levels reached by 24 h. In this series of experiments, the results for 3, 4, 5 and 6 h were below the control average (HC) but within the range of control values (From Moscona *et al.*, 1968).

*RNA synthesis* is obligatory for GS induction in the retina; inhibition of RNA synthesis (with 1-10 $\mu$g/ml Act D) at the time of the addition of the inducer completely prevents the increase in enzyme activity or in the amount of enzyme immunoprecipitable by GS antiserum. These doses of Act D do not stop the uptake of the inducer into the cells and do not immediately stop all protein synthesis in the retina but they inhibit more than 95% of uridine incorporation into RNA (Moscona *et al.*, 1968).

If inhibition of RNA synthesis is delayed till various times after the beginning of induction, continued accumulation of GS becomes increasingly less sensitive to this inhibition; this effect is progressive, in that the longer the delay, the higher is the subsequent increase of GS activity. If complete inhibition of RNA synthesis is delayed till 4-5 h after the beginning of induction, the enzyme continues to accumulate reaching by 24 h levels similar to those in normal induction, or only somewhat lower (Alescio *et al.*, 1970). [There is no "superinduction" of GS by Act D in this case, of the kind described by Tomkins *et al.* (1966) for tyrosine transaminase in hepatoma cell cultures.]

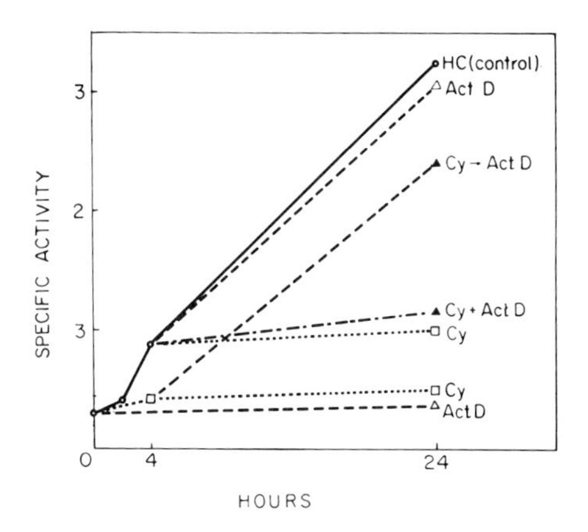


**Figure 11.** Results showing that: (1) after 4 h of induction GS activity in the retina continues to increase although RNA synthesis is halted (10 μg/ml Act D); (2) inhibition of protein synthesis by cycloheximide (2 μg/ml) during the first 4 h of induction does not prevent the subsequent transcription-independent rise in GS activity; (3) protein synthesis is essential for the transcription-independent increase of GS activity after 4 h. All cultures were in medium with inducer. HC (control); cycloheximide (Cy) was added at zero or 4 h, as indicated. Cy → Act D, GS activity in cultures transferred after 4 h in cycloheximide into medium with Act D.

This evidence suggests that during the first hours of induction the inducer brings about accumulation of active RNA templates for GS synthesis; the stability of these templates is indicated by the fact that inhibition of transcription after 4 h of induction does not stop the continuation of enzyme synthesis by the preformed templates (Fig. 10). These conclusions are supported by immunochemical measurements which showed (see Fig. 13) that complete inhibition of RNA synthesis after 4 h of induction halted the rate of enzyme synthesis at the 4 h level (within the period of these measurements; subsequently it declines and stops) while in normal induction the rate of GS synthesis continued to increase, as explained before, presumably due to further addition of active templates (Moscona *et al.*, 1972). Immunoprecipitation measurements have also shown that inhibition of RNA synthesis reduced or stopped the degradation of GS; this suggests that the degradation mechanism is under constant transcriptional control and depends on relatively labile products (Moscona *et al.*, 1972). The inhibition of enzyme degradation by high Act D also accounts for the fact that, in spite of the low and declining rate of enzyme synthesis in this situation, GS may reach by 24 h levels close to those in normally induced tissue, since in the latter enzyme turnover continues (see also Kenney, 1970).

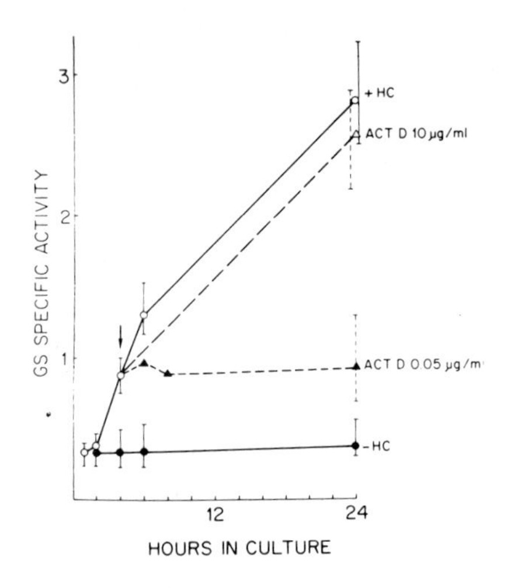


**Figure 12.** Effects of additions of high and low doses of Act D after 4 h of induction on GS specific activity in cultures of embryonic retina with the steroid inducer. HC-induced (hydrocortisone); Arrow indicates time of addition of Act D; vertical lines indicate range of values obtained in different experiments. (From Alescio *et al.*, 1970.)

The inducer can elicit the accumulation of RNA templates for GS without ongoing protein synthesis (Moscona *et al.*, 1968). This was shown (Fig. 11) by adding cycloheximide to retina cultures together with the inducer; after 4 h of incubation cycloheximide was washed out to reinitiate protein synthesis, and at the same time Act D was added to stop further transcription; the cultures were then incubated for 20 h whereupon GS activities were found to have increased significantly. This result demonstrates that: (1) the uninduced cells contain the proteins necessary for transcribing the RNA for GS synthesis; (2) more significantly, it is evident that in this case, the inducer exerts its essential effect not at the level of translational processes, but at the level of transcription, or on post-transcriptive but pre-translational processes.

*The effect of partial inhibition of RNA synthesis.* The above experiments have shown that if after 4 h of induction RNA synthesis in the retina is completely halted, GS continues to be made on the preformed templates and, since enzyme degradation is inhibited, GS continues to accumulate. However, if after 4 h of induction RNA synthesis in the retina is inhibited partially, by about 50-60%, with a low dose of Act D (in the range of 0.05-0.2 μg/ml) GS

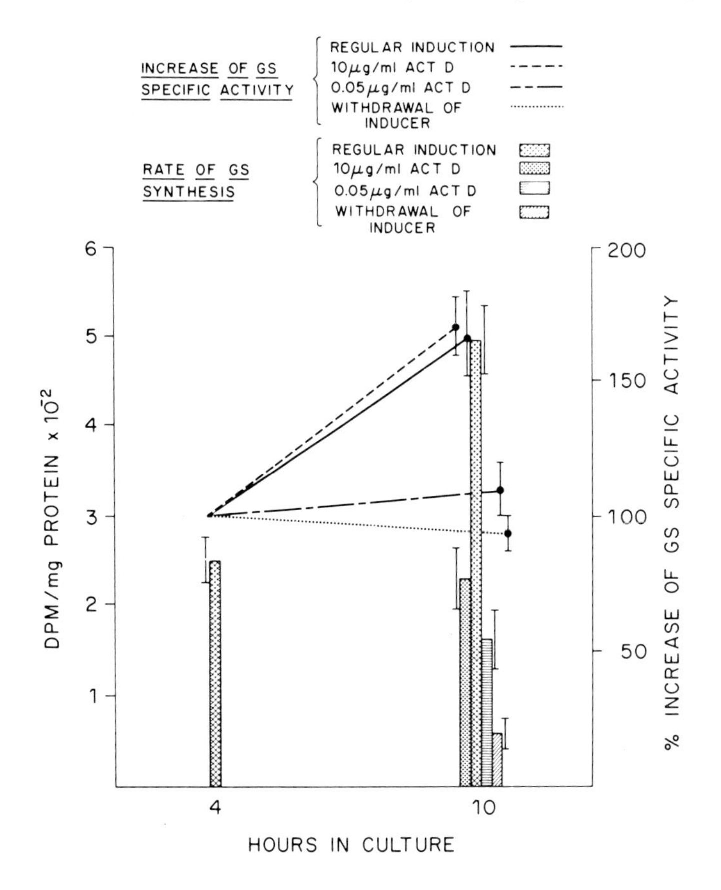


**Figure 13.** Effects of additions of Act D (high and low doses) and of inducer withdrawal, after 4 h of induction on GS specific activity and on the rate of GS synthesis. The rate of synthesis was measured by immunoprecipitation (see legend Figs 6, 7, and text).

accumulation ceases and the enzyme remains at approximately the 4 h level (Fig. 12) (Moscona *et al.*, 1968; Alescio *et al.*, 1970). This low dose of Act D does not immediately stop the formation of templates for GS (if the inducer continues to be supplied), and it does not inhibit preferentially ribosomal RNA synthesis, as in some other cells. Accordingly, the rate of GS synthesis should have continued to go up and the enzyme should have accumulated. However, immunoprecipitation experiments have shown that this low dose of

Act D stops the rate of GS synthesis at the 4 h level, or reduces it somewhat below this (Fig. 13); thus, low Act D acts, in this respect, like the high dose of Act D, in spite of their different effects on total RNA synthesis; but, while the high dose allows GS to increase, the low dose does not (Moscona *et al.*, 1972).

The simplest explanation for the fact that GS does not accumulate in these circumstances is that the low dose of Act D does not stop GS degradation; the 4 h rate of GS synthesis is thus balanced by degradation resulting in a steady level of the enzyme (Moscona *et al.*, 1972). However, the question remains as to why this low dose of Act D prevents the rate of GS synthesis from continuing to increase, or reduces it even though it does not rapidly stop formation of RNA templates for GS. This problem is of considerable interest since it may be related to regulation of template expressivity. For a possible explanation one must turn to hypothetical assumptions based on the available evidence.

## A MODEL FOR THE CONTROL OF GS INDUCTION

The simplest assumption is that the expression of RNA templates for GS is controlled by two relatively labile post-transcriptional regulators: a suppressor and a desuppressor; their formation is differentially sensitive to the low dose of Act D: the desuppressor is inhibited by it, the suppressor is not. This assumption is part of a working model for the regulation of GS induction, suggested first some time ago (Moscona *et al.*, 1968) and schematized in Fig. 14 to include more recent information. Similar models have been suggested also for other cases of enzyme induction (Tomkins, 1969).

The salient feature of our model is that the induction of GS is controlled coordinately at the level of the genome and at the level of expression of RNA templates. The postulated genetic elements are those required for the transcription of mRNA for GS, and genes for the labile suppressor (S) and desuppressor (DS) which regulate post-transcriptively the expression (function; stability) of GS templates. The degradation (turnover) of GS is also dependent on relatively labile products (DG) and requires continuous RNA synthesis. To simplify the scheme in Fig. 14, the mechanism for enzyme degradation is represented only in part 1 of the scheme and its postulated function is described in the text.

In the *non-induced* retina (Fig. 14,1), transcription for GS templates and for the desuppressor is hindered by a labile factor, *r*, derived from the action of R gene. Some mRNA for GS "leaks" in the non-induced retina and provides for the basal rate of enzyme synthesis; this mRNA is rapidly inactivated by the suppressor. The suppressor (S) is relatively labile, but it is continuously transcribed for and made also in the non-induced retina. Like some other regulatory factors, this suppressor may have additional functions in the cell, and its relation to GS could be one of several roles which it fulfills, depending on the state of the cell.

Let us assume that the inducer, or its active derivative (receptor, complex, etc.) has a high binding affinity for *r* and sequesters or inactivates it; this

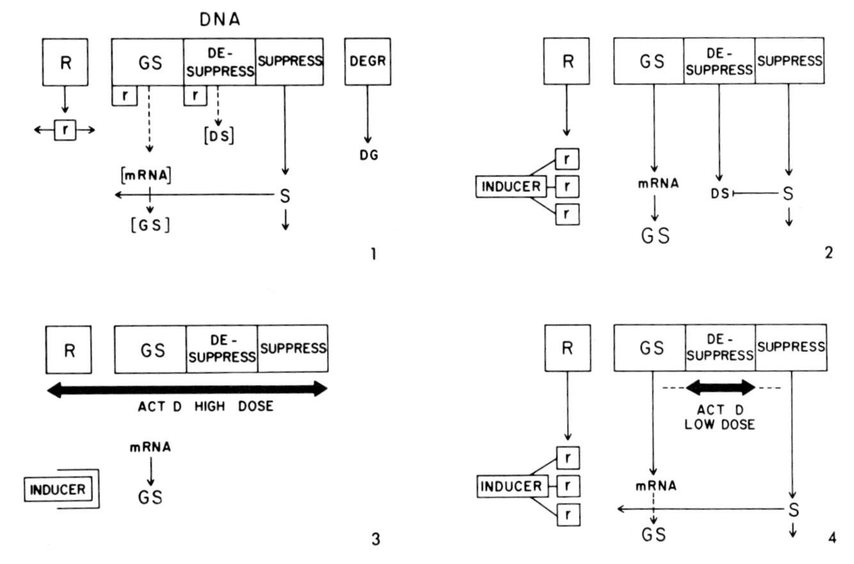


**Figure 14.** A hypothetical scheme for GS induction in embryonic retina. For description, see text. (1) Non-induced state; (2) induced state; (3) complete inhibition of RNA synthesis in induced retina by a high dose of Act D; GS synthesis continues on preformed templates; (4) partial inhibition of RNA synthesis in induced retina with a low dose of Act D; transcription of desuppressor is hindered and the rate of GS synthesis is reduced by the suppressor. For simplicity, the presence of a degrador mechanism is indicated only in (1); in (3), the bracket around the inducer shows that in this situation the inducer is not necessary for continued GS synthesis.

"unblocks" transcription of the genes for GS and for the desuppressor (Fig. 14,2). The desuppressor (DS) neutralizes, or counteracts the effect of the suppressor (S) on GS templates, thereby enabling the accumulation of active templates, and this results in increased enzyme synthesis; although enzyme turnover does not cease, and continuing increase in the rate of GS synthesis results in accumulation of the enzyme.

The suppressor and the desuppressor are relatively labile and their availability depends on RNA synthesis. If after 4 h of induction RNA synthesis is stopped with a high dose of Act D (Fig. 14, 3), these regulators are soon depleted; the formation of templates for GS ceases, but the preformed templates persist and continue to make the enzyme for as long as precursors and components for enzyme synthesis are available. Since, in the absence of RNA synthesis GS degradation stops, the enzyme accumulates, even though it is made at a slow and declining rate.

The situation is different when the low dose of Act D is applied after 4 h of induction: even though transcription of templates for GS does not stop rapidly, the rate of GS synthesis does not increase. Our model (Fig. 14, 4) proposes that low Act D suppresses the rate of GS synthesis because it hinders preferentially transcription for the desuppressor; consequently, the suppressor

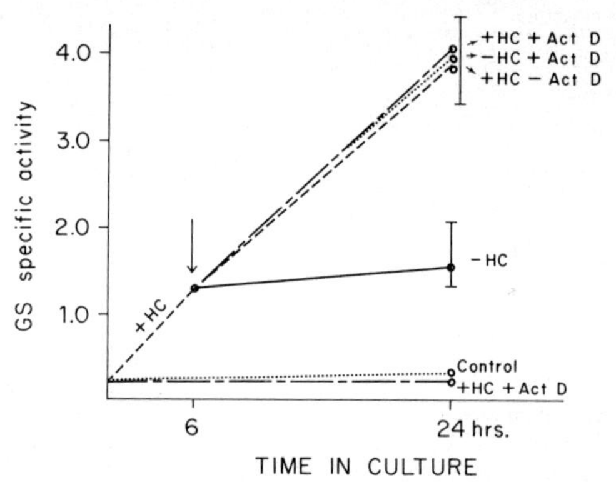


**Figure 15.** The effect of inducer withdrawal (−HC) after 6 hours of induction on GS specific activity in the retina, in the presence and absence of simultaneous addition of Act D (10 μg/ml). 24 h cultures of 12-day embryonic neural retina tissue. Controls were unchanged from zero to 24 h. Arrow indicates time of HC withdrawal and/or addition of Act D. Vertical lines indicate range of values obtained in different experiments.

becomes again available and prevents the rate of GS synthesis from increasing by acting either directly on the processes of translation, or on the stability of the templates. Since enzyme degradation is not stopped by low Act D, it balances the low rate of synthesis resulting in non-accumulation of GS. The preferential sensitivity of the desuppressor to low Act D could conceivably be due to the base composition or frequency of the gene involved in its formation.

The model raises certain testable predictions. Concerning the sequestering of *r* by the inducer, if *r* is made continuously, then, in order to maintain the induced increases in the rate of GS synthesis the inducer must be constantly in supply; its withdrawal from induced retina should make *r* available again, thus causing the genes for GS and for the desuppressor to become again blocked; the resulting depletion of the desuppressor will make the suppressor available and, according to the model, this should reduce sharply the rate of GS synthesis. The experimental findings are consistent with this prediction; immunochemical measurements showed (Moscona *et al.*, 1972) that withdrawal of the inducer after 4 h of induction reduced the rate of GS synthesis eventually almost to base level resulting in cessation of enzyme accumulation (Fig. 13).

The model predicts that this effect should be prevented if, at the time of removal of the inducer, the formation of the suppressor is also stopped by inhibiting all RNA synthesis; GS synthesis should then continue at the 4 h level on the preformed templates and the enzyme should accumulate in the absence of inducer. This prediction is supported by experimental results (Moscona *et al.*, 1972) shown in Figs. 13, 15. It is of interest that comparable results were obtained also in the case of tyrosine transaminase induction in hepatoma cell cultures, suggesting a model for its regulation (Tomkins, 1969) that is largely consistent with the scheme proposed for GS.

A definitive test of the above model requires the availability of mutations which affect GS synthesis. In theory, the effects of gene deletions might be simulated by transcriptional inhibitors with selective actions on the expression of genes involved in GS induction. Conceivably, inhibitors might be found which stop only minimally overall RNA synthesis but inhibit the transcription for GS; others might stop the bulk of transcription but allow that necessary for GS synthesis.

Among agents tested for such effects two show promise for future analysis. Proflavin, at low concentrations, inhibits in the retina approximately 10% of RNA synthesis only, yet it completely prevents the induction of GS; one currently examined possibility is that this is due to a "preferential" inhibition of synthesis of mRNA for GS (Wiens and Moscona, 1972).

Of related interest is the opposite effect of *cytosine arabinoside* in this system; at $7 \times 10^{-3}$ M preparations of this drug stop not only DNA synthesis in the retina, but also block very rapidly more than 80% of RNA synthesis; the remarkable fact is that, at this high concentration this inhibitor causes an "induction" of GS in the complete absence of the steroid inducer (Moscona *et al.*, 1970). Although various regulatory, kinetic and immunological aspects of this induction closely resemble those in steroid induction it remains to be determined whether in both cases a fundamentally similar mechanism is involved. One possibility is that the massive inhibition of RNA synthesis by cytosine arabinoside does not include transcripts for GS synthesis, thus resulting in an apparent "induction" of GS.

Needless to say, these are, at present, highly speculative experimental approaches which, in the framework of the postulated model, will hopefully provide guidelines in the analysis of this and related systems. The proposed model is predicated on the existence in eukaryote cells of regulatory genes for controlling coordinately post-transcriptive and transcriptive processes. The likelihood of this has been accepted since McClintock's (1961) work on maize, and supported by other studies (see: Jacob and Monod, 1963). The discovery of mutations in mice which affect rates of enzyme synthesis (Schimke and Doyle, 1970; Coleman, 1971), as well as evidence from studies on enzyme induction in hepatoma cells (Tomkins, 1969; Kenney, 1970), and on hybrids of hepatoma cells and fibroblasts (Schneider and Weiss, 1971; Thompson and Gelehrter, 1971) also suggest the existence of regulatory mechanisms in mammalian cells entirely consistent with the general features of the model postulated for GS induction in the neural retina.

## ACKNOWLEDGEMENTS

The work reviewed here was supported by research grants HD-01253 from the National Institutes of Health, and GB-6502 from the National Science Foundation. I gratefully acknowledge the collaboration of my colleagues and students in various phases of this research project: M. Moscona, J. L. Hubby, D. L. Kirk, R. Piddington, Y. Shimada, T. Alescio, J. E. Morris, P. K. Sarkar, A. W. Wiens, N. Frenkel, S. A. Garfield and R. E. Jones.

## REFERENCES

Alescio. T. and Moscona, A. A. (1969). *Biochem. biophys. Res. Commun.* **34**, 176.
Alescio, T., Moscona, M. and Moscona, A. A. (1970). *Expl. Cell Res.* **61**, 342.
Chader, G. J. (1971). *Archs Biochem. Biophys.* **144**, 657.
Coleman, D. L. (1971). *Science, N.Y.* **173**, 1245.
Cox, R. P., Elson, N. A., Tu, S. H. and Griffin, M. J. (1971). *J. molec. Biol.* **58**, 197.
Dingman, C. W., Aronow, A., Bunting, S. L., Peackock, A. C., O'Malley, B. W. (1969). *Biochemistry* **8**, 489.
Etzler, M. E. and Moog, F. (1966). *Science, N.Y.* **154**, 1037.
Goldwasser, E. (1966). *In* Current Topics in Developmental Biology (A. A. Moscona and A. Monroy, eds), Vol. I, p. 173. Academic Press, New York and London.
Jacob, F. and Monod, J. (1963). *In* Cytodifferentiation and Molecular Synthesis (M. Locke, ed.) p. 30. Academic Press, New York and London.
Kenney, F. T. (1970). *In* Mammalian Protein Metabolism (H. N. Munro, ed.) p. 131. Academic Press, New York and London.
Kirk, D. L. and Moscona, A. A. (1963). *Devl. Biol.* **8**, 341.
Manchester, K. L. (1970). *In* Mammalian Protein Metabolism (H. N. Munro, ed.) p. 229. Academic Press, New York and London.
McClintock, B. (1961). *Am. Nat.* **95**, 265.
Meister, A. (1968).. *Adv. Enzymol.* **31**, 183.
Mills, E. S. and Topper, Y. J. (1970). *J. Cell. Biol.* **44**, 310.
Moscona, A. A. (1971). *In* Hormones in Development (M. Hamburgh and E. J. W. Barrington, eds) p. 169. Appleton-Century-Crofts, New York.
Moscona, A. A. and Hubby, J. L. (1963). *Devl. Biol.* **7**, **192**.
Moscona, A. A. and Kirk, D. L. (1965). *Science, N.Y.* **148**, 519.
Moscona, A. A. and Piddington, R. (1966). *Biochim. biophys. Acta* **121**, 409.
Moscona, A. A. and Piddington, R. (1967). *Science, N.Y.* **158**, 496.

Moscona, A. A., Moscona, M. and Saenz, N. (1968). *Proc. natn. Acad. Sci. U.S.A.* **61**, 160.
Moscona, A. A., Moscona, M. and Jones, R. E. (1970). *Biochem. biophys. Res. Commun.* **39**, 943.
Moscona, M., Frenkel, N. and Moscona, A. A. (1972). *Devl. Biol.* **28**, 229.
Piddington, R. (1967). *Devl Biol.* **16**, 168.
Piddington, R. (1970). *J. Embryol. exp. Morph.* **23**, 729.
Piddington, R. (1971). *J. exp. Zool.* **177**, 219.
Piddington, R. and Moscona, A. A. (1965). *J. Cell Biol.* **27**, 247.
Piddington, R. and Moscona, A. A. (1967). *Biochim. biophys. Acta* **141**, 429.
Reif-Lehrer, L. (1968). *Biochim. biophys. Acta* **170**, 263.
Reif-Lehrer, L. and Amos, H. (1968). *Biochem. J.* **106**, 425.
Ronzio, R. A., Rowe, W. B., Wilk, S. and Meister, A. (1969). *Biochemistry* **8**, 2670.
Rutter, W. J. and Weber, C. S. (1965). *In* Developmental and Metabolic Control Mechanisms and Neoplasia. Williams and Wilkins, Baltimore.
Salganicoff, L. and deRobertis, E. (1965). *J. Neurochem.* **12**, 287.
Sarkar, P. K. and Moscona, A. A. (1971). *Proc. natn. Acad. Sci. U.S.A.* **68**, 2308.
Sarkar, P. K., Fischman, D. A., Goldwasser, E. and Moscona, A. A. (1972). *J. biol. Chem.* in press.
Schimke, R. T. and Doyle, D. (1970). *A. Rev. Biochem.* **39**, 929.
Schneider, J. A. and Weiss, M. C. (1971). *Proc. natn. Acad. Sci. U.S.A.* **68**, 127.
Shimada, Y., Fischman, D. A. and Moscona, A. A. (1967). *J. Cell Biol.* **35**, 445.
Stephenson, J. R., Axelrod, A. A., McLeod, D. L. and Shreeve, M. M. (1971). *Proc. natn. Acad. Sci. U.S.A.* **68**, 1542.
Sussman, M. (1966). *In* Current Topics in Developmental Biology (A. A. Moscona and A. Monroy, eds), Vol. I, p. 61. Academic Press, New York and London.
Tata, J. R. (1971). *In* Current Topics in Developmental Biology. In press.
Tate, S. S. and Meister, A. (1971). *Proc. natn. Acad. Sci. U.S.A.* **68**, 781.
Tomkins, G. M. (1969). *In* Control of Specific Gene Expression in Mammalian Cells, p. 145. University of Utah Press, Salt Lake City.
Tomkins, G. M., Thompson, E. B., Hayashi, T., Gelehrter, D., Granner, D. and Peterkofsky, B. (1966). *Cold Spring Harb. Symp. quant. Biol.* **31**, 349.
Thompson, E. B. and Gelehrter, T. D. (1971). *Proc. natn. Acad. Sci. U.S.A.* **68**, 2589.
Turkington, R. W. (1968). *In* Current Topics in Developmental Biology (A. A. Moscona and A. Monroy, eds) Vol. 3, p. 199. Academic Press, New York.
Wiens, A. W. and Moscona, A. A. (1972). *Proc. natn. Acad. Sci. U.S.A.* **69**, 1504.
Yamada, T. (1967). *In* Current Topics in Developmental Biology (A. A. Moscona and A. Monroy, eds) Vol. 2, p. 247. Academic Press, New York and London.

FEBS Symposium, Volume 24, 1972, pp. 25-39

# Molecular Hybridization between Labelled 9S Messenger RNA and DNA from the Red Blood Cells of Normal and Anaemic Ducks

J. O. BISHOP

*Department of Genetics, University of Edinburgh, Edinburgh, U.K.*

and

R. E. PEMBERTON

*Department of Biology, Massachusetts Institute of Technology, Cambridge, Massachusetts, U.S.A.*

## INTRODUCTION

The very large amount of DNA in the haploid chromosome complement of higher eukaryotes (Mirsky and Ris, 1951; Vendrely, 1949) has become a matter of increasing interest. At the present time, we have no clear information as to its function. We do not know how much of it is involved directly or indirectly in specifying protein structure, or how much of it is transcribed within the life-span of the organism.

One possibility which must be considered is that structural genes, and perhaps also their controlling sequences, are present in multiple copies. This might come about in two ways. Multiple copies might be present in all cells of the organism (sequence reiteration) or multiplication of each sequence might occur only in those cell lines in which the sequence is utilized (sequence amplification). Precedents for both phenomena are provided by the ribosomal DNA cistrons, which are highly reiterated probably in all organisms and are amplified during oögenesis at least in amphibia and insects (reviewed by Attardi and Amaldi, 1970). The physico-chemical studies of Britten and Kohne (1968) demonstrate the existence of reiterated sequences in the DNA of higher organisms. On the other hand, sequence reiteration is not compatible with classical genetics unless there exists a special mechanism for maintaining uniformity among the reiterated sequence such as the master-slave mechanism postulated by Callan (1967).

If multiple DNA sequences for particular protein sequences do in fact occur, we would expect the phenomenon to be most marked in the case of proteins which are produced in great abundance. The obvious choice is haemoglobin, particularly so because the haemoglobin messenger RNA has been shown to be abundant in immature red blood cells (Marbaix and Burny, 1964). Its identity has been proven by the cell-free synthesis of haemoglobin (Lockard and Lingrel, 1969) and it has been rather well characterized (Chantrenne *et al*, 1967; Labrie, 1969). The haemoglobin messenger RNA of the duck was chosen, because the nucleated circulating red blood cells of ducks continue to synthesize 9S RNA *in vitro* (Scherrer *et al*., 1966; Attardi, *et al*., 1966).

The recently-developed method of DNA-RNA hybridization in vast DNA excess (Melli *et al*., 1971) provides a means of measuring the reiteration frequency of a DNA sequence by hybridizing the total DNA to RNA complementary to that DNA sequence. Here we present the results of hybridizing DNA from the red blood cells of normal and anaemic ducks with labelled 9S messenger RNA.

## MATERIALS AND METHODS

Young Aylesbury ducks were made anaemic by means of phenylhydrazine hydrochloride, blood was withdrawn and the red cells were centrifuged and washed three times with NKM (Borsook *et al*., 1957) and finally with MEMS (Eagle, 1959). The cells were incubated for 8 to 24 h (37°C, 5% $CO_2$ in air) in ten volumes of MEMS supplemented with 10% dialysed anaemic duck serum, glutamine, non-essential amino acids, ferrous ammonium sulphate and either 3 m Ci/ml of $^3$H-uridine (23 Ci/m mole) plus $10^{-4}$ M adenosine, guanosine and cytosine or 3 m Ci/ml of $^3$H-uridine plus 1 m Ci/ml of $^3$H-adenosine (6 Ci/m mole). After incubation the cells were washed three times with NKM and then lysed by osmotic shock as described by Pemberton *et al.* (1971) except that 50 μg/ml dextran sulphate was added. Cell debris was removed by centrifugation and polyribosomes were prepared by centrifugation through 2 M sucrose. In some cases the polyribosome pellet was dissolved in NTE (0.1 M NaCl-0.01 M tris- HCl, pH 7.5 at 0°C, -2 mM EDTA) containing 0.5% SLS and centrifuged for 24 h at 0°C through a 5-20% sucrose gradient (in NTE) in the MSE 6 x 15 ml rotor (Fig. 1). In others, 16S RNP was first prepared by EDTA treatment and sucrose gradient centrifugation essentially as described by Pemberton *et al.* (Fig. 2). The 16S RNP was precipitated with ethanol, dissolved in NTE containing 0.5% SLS and centrifuged through a 5-20% sucrose gradient as described above (Fig. 3). In either case the 9S RNA collected from the 5-20% sucrose gradient was stored frozen at –20°C until use.

DNA was prepared from the nuclei of duck red blood cells lysed by osmotic shock. The nuclei were washed three times in NKM and then lysed in NTE containing 1% SLS. The lysed nuclei were adjusted to 1 M $NaClO_4$ and 0.3 M tris- HCl, pH 8.5 at 2°C, and extracted with phenol and chloroform. After centrifugation the aqueous phase was removed and the combined phenol-chloroform phase and interphase were re-extracted with 1 M $NaClO_4$, 0.3 M tris- HCl, 1% SLS repeatedly until the aqueous phase no longer became viscous. The DNA was then spooled out, after adding an equal volume of

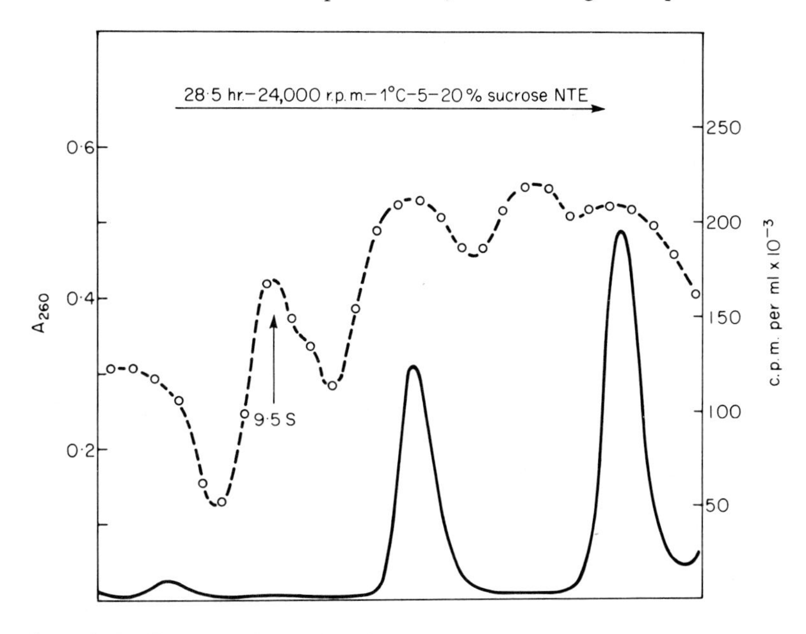


**Figure 1.** Purification of labelled 9S RNA by sucrose-gradient centrifugation of SLS-treated polyribosomes. 28.5 h centrifugation at 24,000 r.p.m. Sedimentation from left to right. The unbroken line shows the $E_{260}$, the broken line the radioactivity.

ethanol, dissolved in 0.1 x SSC, and dialysed against 2 x SSC. Heat-treated ribonuclease was added to 100 μg/ml and the DNA was incubated for 3 h at 37°C. Pronase was then added to 400 μg/ml and after a further 3 h the solution was made 1 M $NaClO_4$, 0.3 M tris- HCl and extracted twice with phenol-chloroform. The aqueous phase was dialysed exhaustively against 2 x SSC, and finally against 0.1 x SSC, passed through a Servall French Press at 50,000 p.s.i. (10-15°C) adjusted to 0.2 M Na-acetate, pH 5, and precipitated with 2 volumes of ethanol. The precipitate was dissolved in 0.3 M NaCl, 0.01 M Na-acetate, pH 5, and passed through a column of Sephadex SP-50. The break through peak was precipitated with 2 volumes of ethanol, dried and dissolved in 0.1 x SSC.

cRNA (RNA synthesized *in vitro* on a template of native duck DNA) was prepared as described previously (Melli *et al.*, 1971). Ribosomal RNA was prepared from a primary explant of duck embryo (15 day) cells labelled for 48 h in MEMS supplemented with duck serum, glutamine, non-essential amino acids and 50 $\mu$ Ci/ml of $^3$H-uridine (23 m Ci/m mole). After washing in NKM the cells were lysed in 1% NaCl, 6% 2-amino salicylate, 1% tri-isopropyl

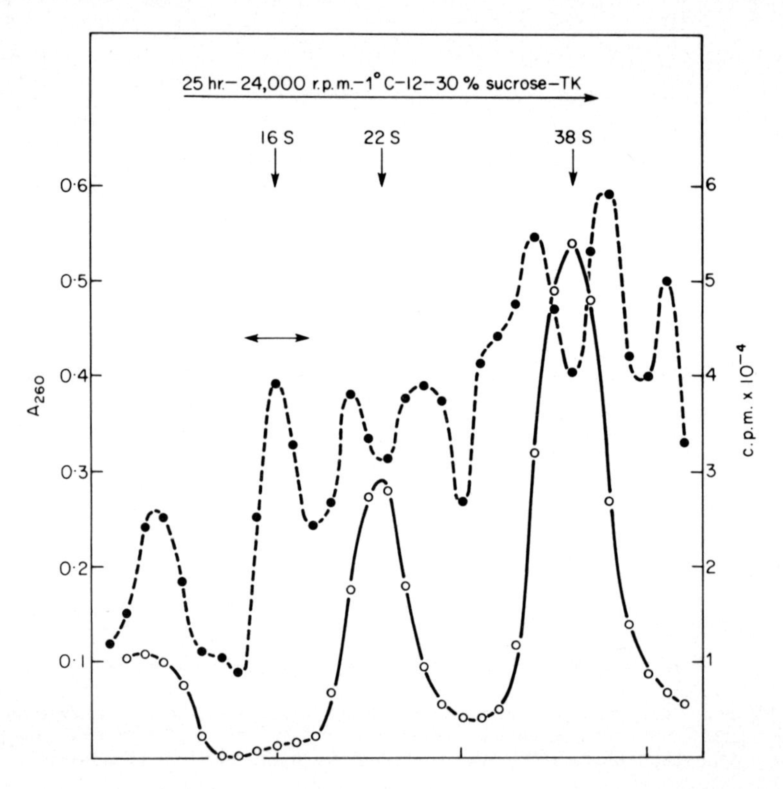


**Figure 2.** Purification of labelled 16S RNP by sucrose-gradient centrifugation of EDTA-treated polyribosomes. 25 h centrifugation at 24,000 r.p.m. Sedimentation from left to right. The unbroken line shows the $E_{260}$, the broken line the radioactivity. Fractions 9-11 were pooled.

naphthalene sulphonate (Parish and Kirby, 1966) and extracted with phenol-cresol-hydroxyquinoline (Parish and Kirby, 1966). The RNA was precipitated with 2 vol. of ethanol, and then reprecipitated repeatedly from 0.2 M Na-acetate, pH 5. Finally it was dissolved in NTE and the ribosomal RNA components were resolved by centrifugation for 20 h at 24,000 r.p.m. and 1°C in the MSE 6 x 15 ml rotor.

DNA-RNA hybridization in vast DNA excess was carried out as described by Melli *et al.* (1971) and Bishop (1972) with the following modifications necessitated by the use of $^3$H-labelled RNA. After ribonuclease treatment the samples were precipitated from 7% TCA in the presence of 100 μg/ml bovine serum albumin for 20 min at 0°C. The precipitates were collected on 2.5 cm diameter glass-fibre discs (Whatman, GF/C) and thoroughly washed with 5%

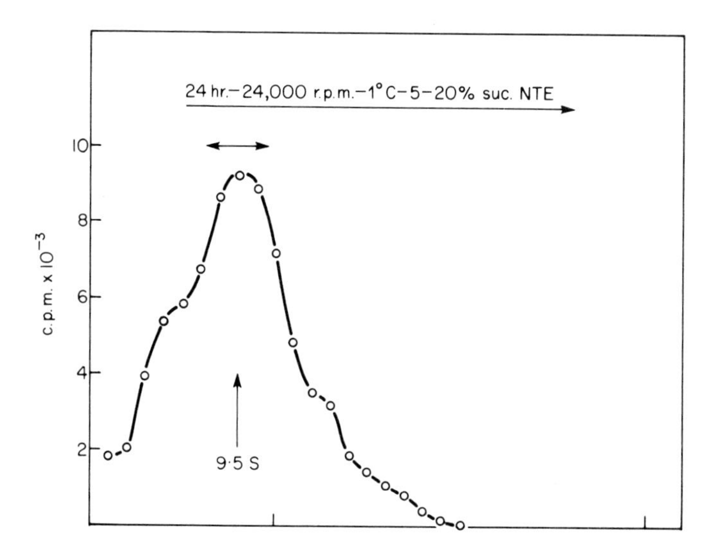


**Figure 3.** Purification of labelled 9S RNA from 16S RNP. 24 h centrifugation at 24,000 r.p.m. Fractions 7-9 were pooled.

TCA. After transfer to counting vials, TCA was removed by placing the vials for 20 minutes at 80°C in a vacuum oven under reduced pressure. When cool, 0.4 ml of ammonia solution (35% $NH_3$) was added to each disc and the vials were left in a ventilated cupboard for 2 hours. 10 ml of counting fluid was then added (7.2 g butyl PBD, 600 ml toluene, 400 ml 2-ethoxyethanol).

The annealing conditions used in these experiments were 2 x SSC at 70°C or (for ribosomal RNA) 75°C.

## RESULTS

### Renaturation of Duck DNA

Renaturation measurements were carried out as previously described (Melli *et al.*, 1971; Bishop, 1971). The DNA was denatured in 0.1 x SSC for 5 minutes at 98-100°C and control samples were immediately diluted into cold 0.1 x

SSC. The remainder was placed at the annealing temperature, and after 30 seconds a concentrated salt solution was added to give a final salt concentration of 2 x SSC (pH 6.4). Samples were withdrawn at different times and diluted with cold solution to a final 0.1 x SSC. The samples are related to each other by the product of initial DNA concentration (Co, moles nucleotide/1) and time (sec). The extinction of each sample at 260 m$\mu$ was then

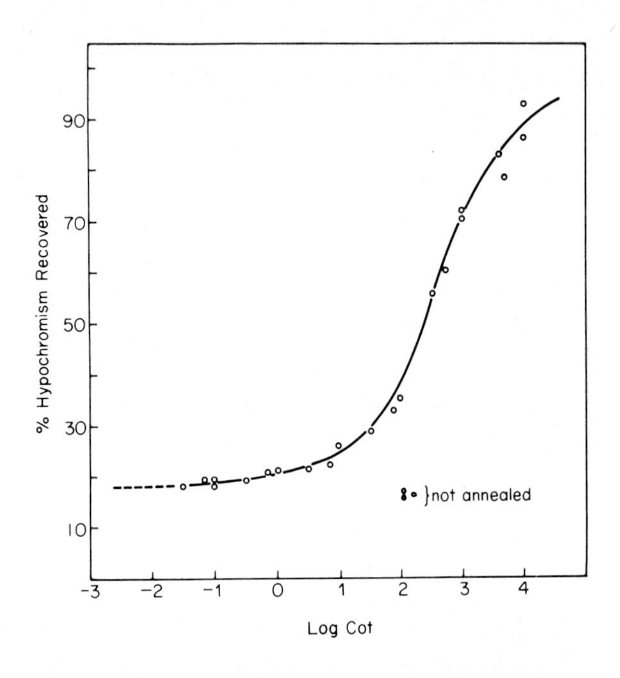


**Figure 4.** Renaturation of sheared duck DNA.

measured first at 50°C, then at 95°C. The difference between these measurements, when compared with that for native DNA, provides a first measure of the proportion of duplex present in the sample (a complete description is given in Bishop, 1972). The renaturation of duck DNA according to this procedure is shown in Fig. 4.

The zero-time sample (Fig. 4) shows a high apparent level of annealing. A similar effect found with bacterial DNA can be attributed to base-stacking in single-stranded DNA (Bishop, 1971), not to the melting of DNA duplex. The hypochromism of the zero-time sample of duck DNA is also due, at least in greater part, to single-strand effects. This may be shown by means of a temperature profile of a zero-time sample (Fig. 5). The profile is very shallow and broad, and there is no sign of an inflection in the region of the DNA Tm (70°C). To obtain an accurate picture, therefore, of the renaturation of the DNA, an adjustment is made for this effect by subtracting the zero-time sample from the others and normalizing. It is convenient to make allowance at the same time for the small percentage of the total DNA which does not appear to react at all. The result is shown in Fig. 6.

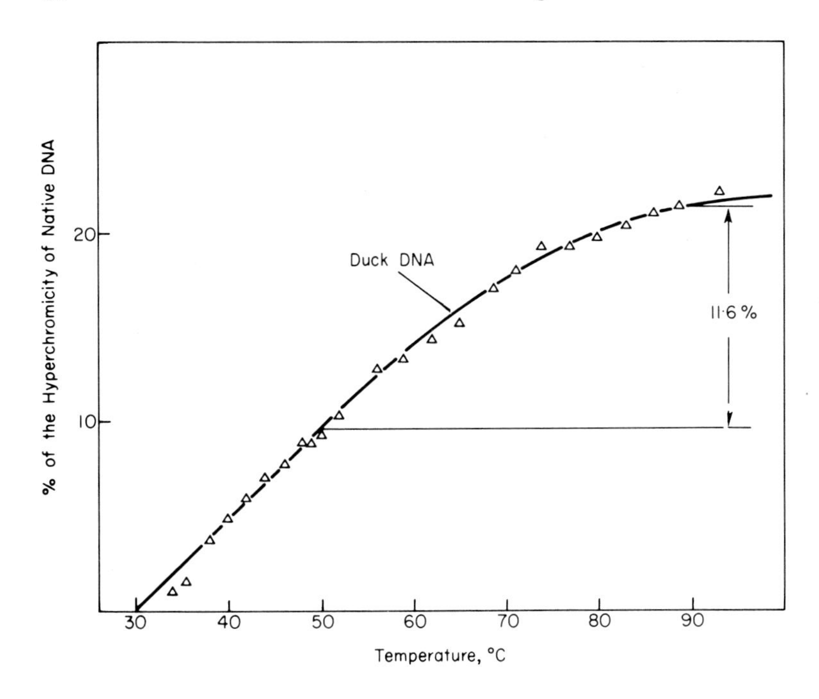


**Figure 5.** Hypochromism of denatured duck DNA.

From similar experiments with DNA of micro-organisms (Bishop, 1972) it is possible to calculate a predicted $Cot^{d}_{1/2}$ for non-reiterated sequences of duck DNA, on the assumption that $Cot^{d}_{1/2}$ is linearly proportional to analytical complexity (Britten and Kohne, 1968; Wetmur and Davidson, 1968). We cannot expect this value to be terribly precise, because an extrapolation of two orders of magnitude is involved. The value, about 600, falls near to the 70% point of the renaturation curve. It seems reasonable to conclude that the bulk of duck DNA consists of sequences reiterated between one and about 10-fold, with the bias probably towards the smaller number. At the other end of the distribution, less than 10% of duck DNA renatures as though composed of sequences reiterated 100 times or more.

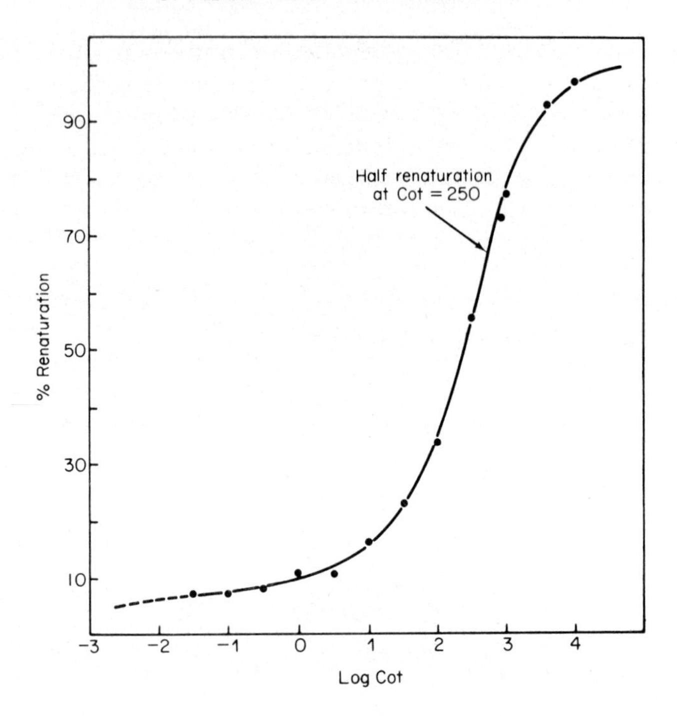


**Figure 6.** Renaturation of sheared duck DNA, corrected for the hypochromism of denatured DNA.

## Hybridization between Duck DNA and cRNA

The hybridization of duck DNA with cRNA in vast DNA excess is shown in Fig. 7. The shape of the hybridization curve is similar to that of the renaturation curve, with hybridization falling behind renaturation, partly because of the difference in the rate-constants of renaturation and hybridization (Melli *et al.*, 1971; Bishop, 1972), partly because of RNA degradation (unpublished experiments). However, nearly 50% of the total cRNA became hybridized at Cot = 10,000, and half of this was observed at a Cot of about 300, which compares with a Cot of 250 for half renaturation of the duck DNA.

Making allowance for the ribonuclease-resistance of zero-time controls (which does not differ significantly from that of heat-denatured cRNA alone), about 4% of the total RNA was hybridized at a Cot of 6-10 (100 or more reiterations). The RNA which hybridizes at low Cot values is 50-60% ribonuclease-sensitive (Bishop, unpublished results).

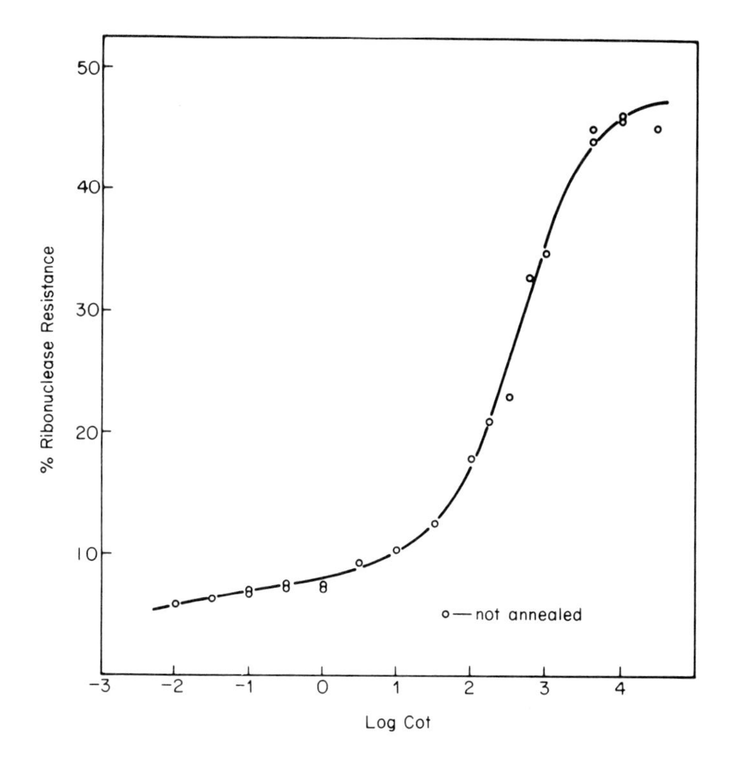


**Figure 7.** Hybridization between duck DNA and cRNA in vast DNA excess.

## Hybridization between Duck DNA and rRNA

The hybridization curve of 18S ribosomal RNA is shown in Fig. 8. Similar results have been obtained using 28S rRNA. The $Cot^{h}_{1/2}$ is about 16, which, by comparison with model systems, represents a 150-fold reiteration frequency for the ribosomal DNA cistrons.

## Hybridization between Duck DNA and 9S Messenger RNA

If there are very few copies of the haemoglobin cistrons per haploid genome, they will make up a very tiny part of the total DNA. For example, five copies of a sequence of molecular weight 300,000 represents 0.00018% of the total DNA. Ideally, hybridization in vast DNA excess requires a ratio of complementary DNA to RNA sequences greater than about 100, in terms of this example an overall DNA/RNA ratio of about $5 \times 10^7$.

In practice, reasonable results can be obtained with lower DNA/RNA ratios. Deviations from ideal kinetics occur, and these have been investigated using ribosomal RNA as a model system (Bishop, 1972). As the complementary sequence ratio falls, the percentage of the RNA hybridized at high Cot values falls and at the same time the *apparent* $Cot_{1/2}$ of the hybridization reaction takes a lower value. The result of failing to obtain vast DNA excess, in other words, will be to overestimate slightly the reiteration frequency of the complementary DNA sequences.

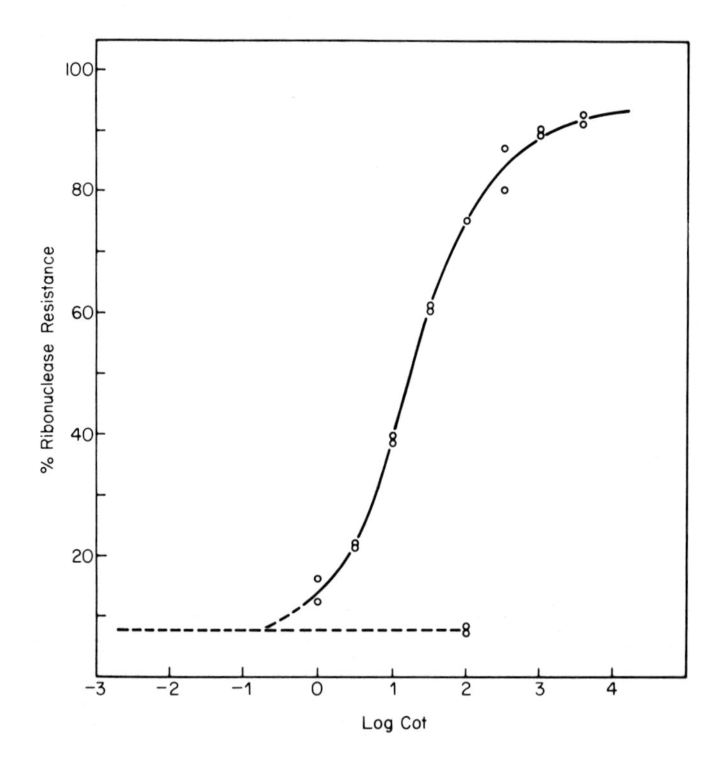


**Figure 8.** Hybridization between duck DNA and 18S rRNA at a DNA/RNA complementary sequence ratio of about 100.

If the number of haemoglobin DNA cistrons is small, a very high specific-activity messenger RNA is required if we are to obtain DNA excess at all. It is presently impossible to obtain a large amount of high specific-activity 9S RNA, mainly because of the cost of high specific-activity uridine. Consequently, specific-activity measurements are not very accurate. The amount of labelled 9S RNA obtained in each preparation was estimated from the slight deflection of the recorder which occurred as the final sucrose gradient passed through an Isco U.V. analyser. Two preparations were used in the experiments described here. One, obtained by centrifuging SLS-treated polyribosomes through a sucrose density gradient (Fig. 1) had a specific activity of about $3 \times 10^5$ cpm/$\mu$g. The other, obtained after isolation of 16S RNP (Figs 2 and 3) had a specific activity of about $1.5 \times 10^5$ cpm/$\mu$g.

Results of hybridizing these preparations are shown in Figs 9 and 10. The apparent $Cot^h_{1/2}$ in these experiments was 250-300. By comparison with model systems, this represents a reiteration frequency of about 10 for the

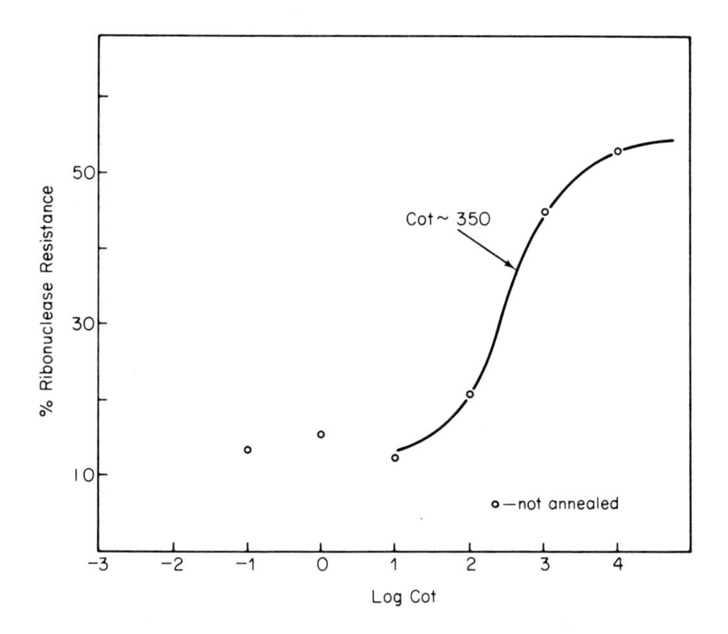


**Figure 9.** Hybridization between duck DNA and 9S RNA from SLS-treated polyribosomes.

complementary DNA sequences. The highest observed levels of hybridization were about 40 and 30%, rather than the 50-60% observed with cRNA or HnRNA, probably because the DNA/RNA ratio is lower in each case than that required for truly vast DNA excess. Because of this, it is probably fair to say that the reiteration frequency of the complementary DNA cistrons is 5 or less.

It should be noted that Figs 9 and 10 show no trace of a component hybridizing at a Cot of 16. In other words, none of the labelled 9S RNA hybridizes in the same way as ribosomal RNA.

### Comparison between DNA from Red Blood Cells of Normal and Anaemic Ducks

The experiments with DNA from red blood cells of normal and anaemic ducks (Fig. 10) show no obvious difference in the hybridization pattern. Similar results have been observed with liver DNA. It may be concluded that there is no major amplification of the haemoglobin DNA cistrons in the circulating red blood cells of the anaemic duck. Since these are actively synthesizing 9S RNA, it seems unlikely that there is amplification at earlier stages of maturation. This possibility, however, is not completely ruled out by these experiments.

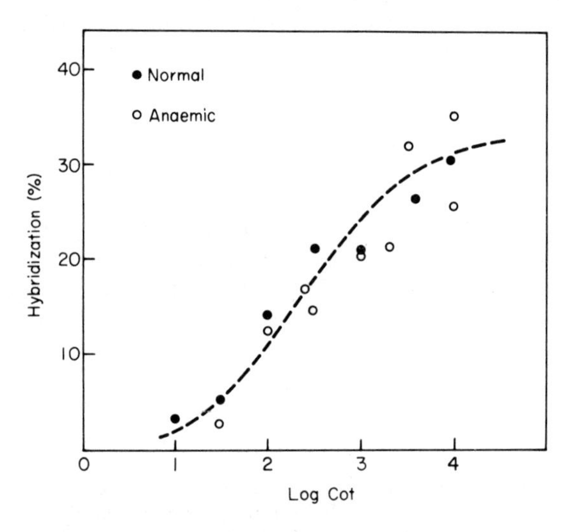


**Figure 10.** Hybridization between duck DNA and 9S RNA from 16S RNP. DNA was isolated from red blood cells obtained from normal (•) or anaemic (○) ducks. Non-annealed control values have been subtracted.

### Dilution of Labelled with Unlabelled 9S RNA

By showing that the 9S RNA isolated from immature duck red blood cells induces the synthesis of duck haemoglobin in a rabbit cell-free system, Pemberton *et al.* (1971) have proven that this RNA contains a proportion at least of haemoglobin messenger RNA. It also contains a proportion (approximately 50%) of ribosomal RNA degradation products (Bishop, unpublished).

The hybridization experiments shown in Figs 9 and 10 show that the 9S RNA which becomes labelled *in vitro* does not contain any labelled ribosomal RNA sequences. This observation is reinforced by unpublished hybridization experiments which show that the cells used in these experiments do not synthesize ribosomal RNA to a significant extent.

The unlabelled 9S RNA, which stimulates haemoglobin synthesis, competes with the labelled 9S RNA in hybridization experiments (Fig. 11); but it is the quantitative aspects of this experiment which are most important.

Competition experiments based on the vast DNA excess hybridization system depend upon breaking the basic rule of the system in a systematic way. As the DNA/RNA (complementary sequence) ratio falls from 100 to 0.1,

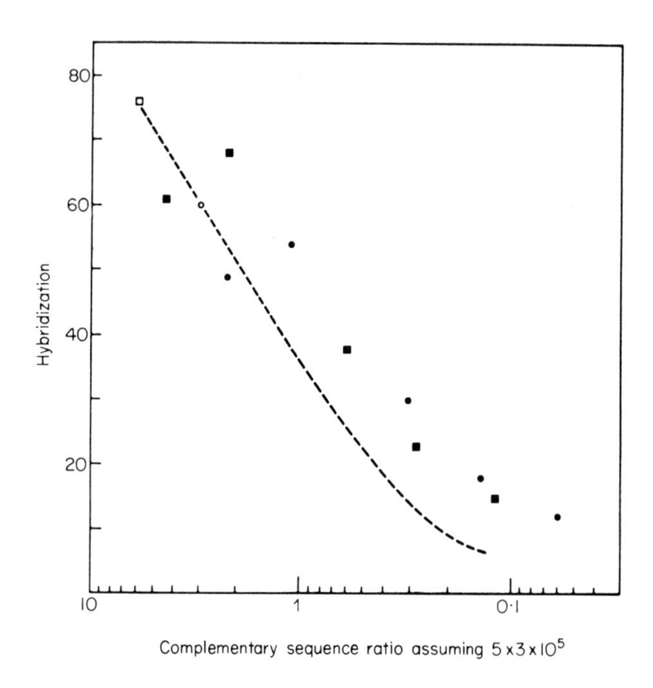


**Figure 11.** Competition between labelled and unlabelled 9S RNA. Two experiments are shown. The control values (○,□) are set (on the basis of the measured specific activity of the RNA) to lie on the model competition curve (broken line) determined by means of experiments with 18S ribosomal RNA.

the amount of RNA becoming hybridized at the highest Cot values falls to 10% of the highest amount (unpublished model experiments with duck ribosomal RNA). Comparison with the model data allows evaluation of other situations. The data in Fig. 11 are plotted on the basis of two assumptions: (1) that duck DNA contains five copies of a sequence of molecular weight 300,000 complementary to 9S RNA; (2) that unlabelled 9S RNA contains only that sequence. The results of the model experiments with ribosomal RNA are represented by the broken line. There is a discrepancy of a factor of approximately three. Most of this is accounted for by the rRNA contamination of the unlabelled 9S RNA. These results show quite unequivocally that the labelled 9S RNA is a relatively short sequence, and that duck DNA contains about five copies of a sequence complementary to it.

## DISCUSSION

In the first place, these results show that labelled 9S RNA isolated from polyribosomes of immature duck red blood cells is complementary to a DNA sequence which is reiterated about 5-fold per haploid chromosome complement. Direct evidence that the labelled 9S RNA is haemoglobin messenger RNA is lacking. However, indirect evidence suggests that it probably is. The hybridization results show that it is not ribosomal RNA, because the 9S RNA hybridizes at much higher Cot values. Secondly, it is diluted out by unlabelled 9S RNA which is known to contain haemoglobin messenger RNA. The quantitative evaluation of the dilution experiments suggests that the complementary RNA and DNA sequences have a total sequence length approximately twice the size of the better characterized haemoglobin messenger RNA of the rabbit (Labrie, 1969; Gaskill and Kabat, 1971). The precision of the method in this case is probably not sufficient to warrant a literal acceptance of this result. The important point is that the labelled sequence is shown to be very short.

Haemoglobin makes up about 90% of the protein synthesized by the red blood cell (Scherrer *et al.*, 1966). The nucleated red blood cell is therefore a likely place to find gene amplification if it does in fact occur in the somatic cells of vertebrates. Our observation that the cistrons complementary to 9S RNA are equally reiterated in mature blood cells and in cells actively synthesizing 9S RNA may be taken as preliminary evidence that gene amplification does not occur in vertebrate somatic cells.

## ACKNOWLEDGEMENTS

This work was supported by the British Science Research Council and the American National Science Foundation. We are very grateful for the excellent technical assistance of Miss Ray Murray and Miss Melville Richardson.

## REFERENCES

Attardi, G. and Amaldi, F. (1970). *A. Rev. Biochem.* **39**, 183.

Attardi, G., Parnas, H., Hwang, M-I.H. and Attardi, B. (1966). *J. molec. Biol.* **20**, 145.

Bishop, J. O. (1972). *Biochem. J.*, **126**, 171.

Borsook, H., Fischer, E. H. and Keighley, G. (1957). *J. Biol. Chem.* **229**, 1059.
Britten, R. J. and Kohne, D. E. (1968). *Science, N.Y.* **161**, 529.
Callan, H. G. (1967). *J. Cell Sci.* **2**, 1.
Chantrenne, H., Burny, A. and Marbaix, G. (1967). *Prog. nucl. Acid Res. Mol. Biol.* **7**, 173.
Eagle, H. (1959). *Science, N.Y.* **130**, 432.
Gaskill, P. and Kabat, D. (1971). *Proc. natn. Acad. Sci. U.S.A.*, **68**, 72.
Labrie, F. (1969). *Nature, Lond.* **221**, 1217.
Lockard, R. E. and Lingrel, J. B. (1969). *Biochem. biophys. Res. Commun.* **37**, 204.
Marbaix, G. and Burny, A. (1964). *Biochem. biophys. Res. Commun.* **16**, 522.
Melli, M., Whitfield, C., Rao, K. V., Richardson, M. and Bishop, J. O. (1971). *Nature Lond.* **231**, 8.
Mirsky, A. E. and Ris, H. (1951). *J. gen. Physiol.* **34**, 451.
Parish, J. H. and Kirby, K. S. (1966). *Biochem. biophys. Acta,* **129**, 554.
Pemberton, R. E., Housman, D., Lodish, H. F. and Baglioni, C. (1971). *Nature, Lond.* in press.
Scherrer, K., Marcaud, L., Zajdela, F., Breckenridge, B. and Gros, F. (1966). *Bull. Soc. Chim. biol.,* **48**, 1037.
Vendrely, R. (1949). *C.r. Séanc Soc. Biol.*, **143**, 1386.
Wetmur, J. G., and Davidson, N. (1968). *J. molec. Biol.,* **31**, 349.

FEBS Symposium, Volume 24, 1972, pp. 41-53

# Cell Cycle-dependent Events During Myogenesis, Neurogenesis, and Erythrogenesis[1]

HOWARD HOLTZER, HAROLD WEINTRAUB and JUDITH BIEHL

*Department of Anatomy, School of Medicine,*
*University of Pennsylvania, Philadelphia, Pennsylvania, U.S.A.*

Problems in cell physiology are often mistaken for problems in cell differentiation. Obviously the study of the way in which polyribosomes function during myosin synthesis in muscle cells or during hemoglobin synthesis in erythroblasts is of great interest. But such studies do not directly illuminate the regulatory mechanisms initiating the first steps in the differentiation of the myogenic or erythrogenic cell lineages. To understand how *only* muscle or *only* red blood cells synthesize myosin *or* hemoglobin it is necessary to study those dimly understood processes structuring the cytoplasmic-nuclear interactions that dictate what kinds of polyribosomes a given cell *may* assemble. Learning more about the product will not enhance understanding of how the cellular machine was assembled.

By the time a myoblast synthesizes its first molecule of myosin or an erythroblast synthesizes its first molecule of hemoglobin the major genetic decisions determining what kinds of molecules these cells would synthesize have long since been made. These decisions were made in part in the mother cell, in part in the grand-mother cell and in part in even earlier ancestral cells. This point of view generates the following questions: How many sequential decisions have to be made between the zygote and the first cell to synthesize myosin or hemoglobin? How does a succession of such decisions preclude the possibility that cells in the myogenic lineage synthesize hemoglobin and that cells in the erythrogenic lineage synthesize myosin? Lastly, what is the role of the cell cycle in initiating these genetic decisions that change one generation of precursor cells into a later more mature generation within an evolving lineage?

The presumptive myoblast is the mother cell of the myoblast. The presumptive myoblast, unlike the myoblast, does not synthesize myosin,

[1] This work was supported by grants from the U.S.P.H.S. (HD-00189), the National Science Foundation (GB-5047x), and the American Cancer Society (VC-45).

tropomyosin, myoglobin or any other terminal luxury molecule associated with mature muscle; yet by any criterion the presumptive myoblast is a committed, covertly differentiated myogenic cell (Holtzer, 1970a,b, Holtzer *et al.*, 1972). The hematocytoblast is the mother cell of the erythroblast that will initate the synthesis of hemoglobin. The hematocytoblast does not synthesize hemoglobin, yet is a covertly differentiated hematogenic cell (Holtzer, 1970b; Weintraub *et al.*, 1971). One cell cycle or cell generation separates the presumptive myoblast or hematocytoblast from the myoblast or erythroblast. What then does "cell maturation" mean in terms of the time from the last mitosis and in terms of cell generations? In this report several experiments are described that are consistent with the following propositions: (1) There are two classes of cell cycles, "quantal" and "proliferative", and that (2) Movement of cells from one compartment into the next compartment within a cell lineage depends upon a "quantal" cell cycle, whereas increasing the numbers of cells within a compartment depends on variable numbers of "proliferative" cell cycles. Quantal cell cycles are cell cycles that yield one or two daughter cells with capacities different from those of the mother cell; proliferative cell cycles yield two daughters with synthetic repertoires identical to those of the mother cell (Holtzer, 1970b; Weintraub *et al.*, 1971; Holtzer *et al.*, 1972).

To learn more about ancestral cells we have been following the effects of BrUdR in the DNA of cells in different generations of defined lineages. Earlier (Stockdate *et al.*, 1964; Okazaki and Holtzer, 1965; Abbott and Holtzer, 1968; Lasher and Cahn, 1969; Shulte-Holthausen *et al.*, 1969; Bischoff and Holtzer, 1970; Coleman and Colman, 1970; Bischoff, 1971; Mayne *et al.*, 1971; Holtzer *et al.*, 1972b) work stressed the following: (1) BrUdR incorporated into replicating myogenic, chondrogenic or amnion cells does not behave as a conventional mutagen, for it affects 100% of the cells; (2) It has no obvious effect on the cell incorporating the analogue but blocks the initiation of the synthesis of myosin, chondroitin sulfate, or hyaluronic acid in the succeeding generation of cells; (3) At low concentrations the analogue inhibits the synthesis of some terminal luxury molecules to a much greater extent than the synthesis of those essential molecules required for cell division and general growth; (4) When BrUdR-suppressed myogenic, chondrogenic, or amnion cells are transferred to normal medium, progeny eventually emerge that differentiate into normal muscle, cartilage, or amnion cells, demonstrating that the analogue does not cancel the genetic commitment of these types of suppressed cells to a particular lineage.

## MYOGENESIS

Myoblasts are the daughters of replicating presumptive myoblasts. They are post-mitotic, mononucleated cells that synthesize myosin, actin, and tropomyosin; they have the capacity to fuse with other myogenic cells to form multinucleated myotubes (Holtzer, 1970a; Holtzer and Sanger, 1972). If BrUdR is incorporated into presumptive myoblasts, the resulting daughter cells do not develop into post-mitotic myoblasts. When BrUdR-suppressed presumptive myoblasts are transferred to normal medium and allowed to undergo several rounds of DNA synthesis some of the 4th and 5th generation cells behave as normal myoblasts (Bischoff and Holtzer, 1970).

Myoblasts are the terminal generation and presumptive myoblasts the penultimate generation in the myogenic lineage. Progenitors to presumptive myoblasts have been termed myogenic beta cells (Holtzer, 1970a; Holtzer *et al.*, 1972a). Recently BrUdR has been introduced into the DNA of replicating myogenic beta cells. Somites from stage 14–16 chick embryos were used. These somites consist of approximately $2 \times 10^4$ cells each. Roughly 70% belong to the myogenic lineage and of these approximately 3% are post-mitotic myoblasts; the remainder of the cells belong to the chondrogenic-fibrogenic lineage. Cell suspensions of these somites are rich in myogenic beta cells. When these cells are grown for 4 days *in vitro* and then sub-cultured for another 4 days, they yield large numbers of spontaneously contracting myotubes and myoblasts. When similar somite cells are grown in BrUdR for 4 days and then sub-cultured they do not develop into myoblasts or myotubes. BrUdR-suppressed myogenic beta cells have been repeatedly sub-cultured in normal medium. However, their progeny, even after many generations, do not yield myoblasts (see also Abbott *et al.*, 1972; Mayne *et al.*, 1972).

In summary: Incorporation of BrUdR into myogenic beta cells blocks these cells from advancing into the next state in the myogenic lineage—that is, from forming by way of a quantal cell cycle daughter presumptive myoblasts. The intriguing finding is that the progeny of BrUdr-suppressed myogenic beta cells replicating in normal medium do not *automatically* yield presumptive myoblasts. The simplest interpretation of these observations is that our tissue culture conditions do not permit large numbers of myogenic beta cells to undergo that particular quantal cell cycle that leads to presumptive myoblasts; on the other hand these cells can undergo large numbers of proliferative cell cycles. As sizeable numbers of these arrested myogenic beta cells can be collected, we hope to characterize their biochemistry more fully.

## DIFFERENTIATION OF RETINA CELLS

The developing retina in the 2-day chick embryo consists of two simple epithelia. One epithelium constitutes the precursor to rods, cones, ganglion cells, and amocrine cells; the other the precursor to the melanin-synthesizing cells of the pigmented retina. If BrUdR blocks the emergence of successive phenotypes in a lineage, then it should be possible to accumulate large numbers of arrested retinal-precursor cells.

Eyes from chick embryos in stages 12–14, 14–16 or 16–18 were grown as organ cultures on millipore rafts for 5 days. The major cytological distinctions between stage 12 and stage 18 retinas are the greater numbers of cells and the incipient psuedo-stratification in the older epithelium. Virtually all cells in both epithelia are in the division cycle. With conventional microscopy, the cells in the neural retina and future pigmented retina are indistinguishable.

After 2 days in organ culture the eyes from stage 16–18 embryos displayed dozens of pigmented cells and by the 5th day several thousand heavily melanized cells had accumulated. These pigment cells were conspicuous in the living cultures under the dissecting microscope.

Tables 1–3 summarize the extent of pigmentation in control and BrUdR-treated eyes. The suppression of pigmentation is evident. Note also that at lower doses of the analogue the total number of cells in the

**Table 1.** Eye explants cultured on millipore rafts for 5 days and then scored as positive or negative for retinal pigment cells.

| | Controls | | BrUdR-treated | |
|---|---|---|---|---|
| | Positive | Negative | Positive | Negative |
| Stage 12–16 | 23 | 0 | 2 | 21 |
| Stage 14–16 | 12 | 0 | 14 | 22 |
| Stage 16–18 | 40 | 0 | 13 | 35 |

**Table 2.** Cell counts on organ cultures of eyes after 4 days in culture. Twelve eyes were used in each category and the figures are the average numbers of cells in each eye.

| | Control | 1 μmg BrUdR/ml | 10 μgm BrUdR/ml | 20 μgm BrUdR/ml |
|---|---|---|---|---|
| Stage 12–14 | $2 \times 10^6$ | $1.7 \times 10^6$ | $1 \times 10^6$ | $0.5 \times 10^6$ |
| Stage 14–16 | $1 \times 10^6$ | $1.0 \times 10^6$ | $1 \times 10^6$ | $0.8 \times 10^6$ |
| Stage 16–18 | $1.5 \times 10^6$ | $1.5 \times 10^6$ | $1.3 \times 10^6$ | $1.0 \times 10^6$ |

**Table 3.** Semi-quantitative evaluation of the degree of pigmentation in eye explants.

| Stage 16–18 | Neg. | 1+ | 2+ | 3+ | 4+ |
|---|---|---|---|---|---|
| Controls | 0 | 0 | 4 | 6 | 10 |
| BrUdR-treated | 13 | 11 | 2 | 0 | 0 |

BrUdR-treated eyes is not greatly reduced. Histological inspection of the BrUdR-treated eyes revealed even more striking suppression of differentiation of the neutral retina.

Additional experiments were performed on these organ cultured eyes. After 4 days in organ culture the control eyes from stages 12–14 were trypsinized and then plated at low-to-moderate densities in Petri dishes. The object of these experiments was to permit further replication and cloning of the various kinds of retinal cells. Cell suspensions prepared from the organ-cultured normal retinas after 4 days of growth displayed islands of pigment cells, islands of epithelial cells, and two classes of nerve cells that grew on the epithelial islands. Eyes grown for 4 days in organ culture in BrUdR and then treated in the same way displayed only islands of epithelial cells. Though the epithelial cells derived from the BrUdR-treated cells replicated many times in normal medium they did not give rise to recognizable pigment or nerve cells. Experiments to be reported in detail elsewhere indicated that every cell incorporated $^3$H-Tdr during the organ culture period and that many cells in the secondary or tertiary cultures represented 5th or 6th generation cells.

These results are interpreted as follows: Incorporation of BrUdR into the DNA of precursor retinal epithelial cells blocks these cells from producing progeny more advanced than the mother cells. Normally 2nd, 3rd or 4th generation progeny of stage-18 presumptive pigment cells differentiates into recognizable pigment cells, whilst 4th, 5th, 6th and later generations of retinal epithelial cells differentiate into recognizable nerve cells. If, however, these early classes of precursors incorporate BrUdR, they do not yield overt pigment or nerve cells even though they undergo many cell cycles in normal medium. They behave precisely as did the myogenic beta cells. We propose that by interfering, not with the total numbers of cell cycles, but with specific quantal cell cycles that only occur in certain microenvironments, the step-wise vectoral progression through the various lineages is blocked.

## ERYTHROPOESIS

What might become an important clue as to the mode of action of BrUdR are its strikingly different effects at different points along the erythropoietic cell lineage. Miura and Wilt (1971) and Wenk (1971) have shown that in cultures of chick hematocytoblasts BrUdR inhibits the emergence of 1st generation erythroblasts, the first cells which synthesize Hb. We have confirmed their results and extended them to the *in vivo* situation. Here the primitive line of erythroblasts begins to appear at 35 h of incubation. The transition from hematocytoblast to Hb-producing erythroblasts occurs over a 24 hour period between 35 h and the second day. Each hematocytoblast is the precursor for 7 subsequent generations of erythroblasts. There are several notable changes in these cells from generation to generation as they mature. Morphologically they become smaller, less basophilic, their nuclei condense, and the ratio of nucleus to cytoplasm decreases. There is a gradual decrease in RNA synthesis, most prominent between the 5th and 6th generations of erythroblasts. Each generation synthesizes characteristic amounts of Hb, which in turn are parcelled to daughter cells of the next generation (Fig. 1). Electrophoretically, there are two major types of Hb as well as two minor types. They are synthesized coordinately throughout the 7 generations, that is, in the same ratios to each other. There is a decrease in the rate of Hb synthesis between the 6th and 7th divisions as well as a lengthening of the cell cycle from three 10 h cycles to two 17 h cell cycles to a final cell cycle of 28 h (Campbell *et al.*, 1971; Hagopian and Ingram, 1971; Hagopian *et al.*, 1972).

By injecting BrUdR into eggs of varying ages, it is possible to monitor the effects of the analogue at various times during this rigorously defined sequence of erythropoietic changes. Figure 2 shows a point of transition before which Hb synthesis is inhibited by BrUdR (50 $\mu$gm/ml) and after which BrUdR has no effect. Where 20%–30% substitution blocks the emergence of myoblasts from presumptive myoblasts and erythroblasts from hematocytoblasts, 80% substitution over 3–4 generations has no effect on red cells already making Hb.

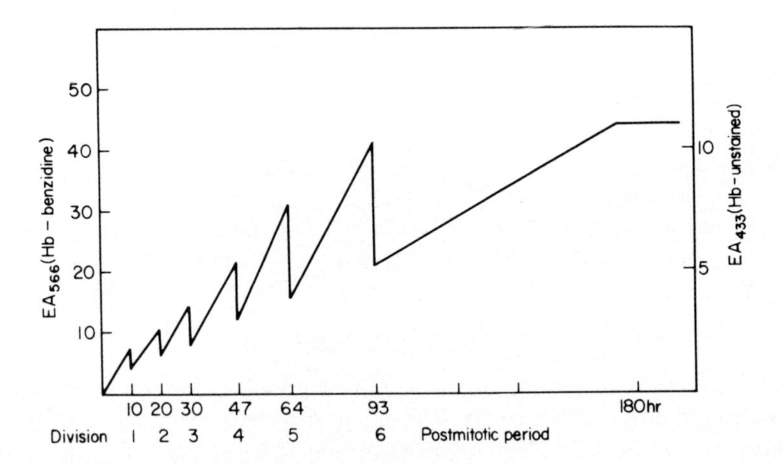


**Figure 1.** Hemoglobin content of primitive erythroblasts was determined cytophotometrically on isolated metaphase-arrested cells stained with benzidine; this ensured reliable measurements of hemoglobin both in terms of geometry and of eliminated variations related to phase of the mitotic cycle. The associated cell cycle length was calculated from the kinetics of $^3$H-TdR labeling. These two types of measurements were obtained for various generations during the maturation of the circulating red blood cells. The compiled data are summarized above. The slope of the lines during the interphase period represents an average rate obtained by dividing the net amount of hemoglobin made in a particular cell cycle by the length of that cell cycle (Holtzer, 1970b).

If the analogue is administered at 45 h of incubation when only half of the hematocytoblast population has moved into the 1st or 2nd generation erythroblast compartment, there is an inhibition of some 40% in the amount of Hb *per embryo* on day 3. Cytophotometric examination of the individual erythroblasts in these embryos revealed a normal amount of Hb per cell. With respect to the parameter "susceptibility to BrUdR", it is possible to isolate two types of divisions during erythropoiesis. Thus, as measured by its ability

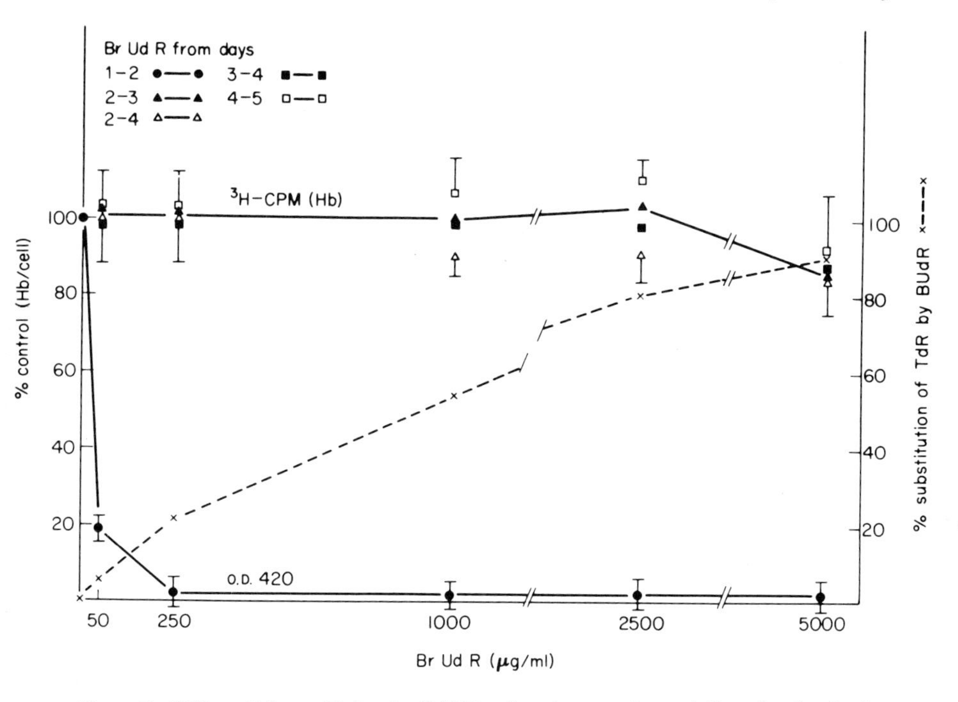


**Figure 2.** Differential sensitivity to BrUdR of various erythropoietic cells. 3 ml of a BrUdR solution was introduced into embryos of different ages. The interrupted line indicates the percentage substitution of BrU for thymine at the various concentrations of BrUdR. This was determined by equilibrium centrifugation on CsCl. The continuous lines indicate the amount of Hb as a percentage of controls for exposure over the periods indicated by the symbols on the figure. This was measured in either of two ways. For incubations between 1 and 2 days, it was sufficient to measure the amount of Hb per embryo (5 embryos). For older embryos, Hb/embryo was measured as was Hb/cell. The latter was done cytophotometrically (Campbell *et al.*, 1971). In addition, the rate of Hb synthesis per cell was determined by incubating 5 x $10^6$ cells in medium F10 (Weintraub *et al.*, 1971) with 5 μc/ml $^3$H-leucine. After one hour, the cells were washed three times in saline and lysed in 0.3% saponin in saline. Nuclei were then pelleted and the supernatant passed through Sephadex G-25 equilibrated in 0.01 M sodium phosphate buffer, pH 6.3. The protein-containing eluate was then passed over CMC equilibrated with the same buffer. Under these conditions, Hb is bound to the column and the majority of the non-Hb protein passes through. Hb is then removed by raising the pH to 9.1 and counted in a scintillation counter.

to inhibit Hb synthesis, BrUdR defines the quantal division between hematocytoblast and erythroblast and distinguishes it from the six subsequent proliferative divisions. This point is further emphasized by our observations that none of the previously mentioned changes that occur during red cell development is affected by BrUdR, nor are there any striking differences (±10%) in the synthesis of histones, acidic nuclear proteins, or soluble non-Hb proteins, as they are displayed in double-label experiments on SDS-polyacrylamide gels.

The transition between sensitive hematocytoblast and resistant erythroblasts appears to reflect the institution of a "program" that is BrUdR-resistant. What might be the biochemical basis for the postulated programming events? One possibility is that the mRNA for HB is stable. However, two inhibitors of RNA synthesis, actinomycin-D which also inhibits DNA synthesis, and cordecypin which does not inhibit DNA synthesis, lead to a prompt inhibition of Hb synthesis. It is possible that the inhibition observed results from the inhibition of some step common to all protein synthesis. That this is probably not the case is indicated by the following: (1) Doses of actinomycin-D reported to inhibit rRNA synthesis preferentially have little effect on the synthesis of HB. (2) Different protein fractions show a varying degree of inhibition by actinomycin-D. Acrylamide-gel analysis of the soluble proteins show much variation in the degree of inhibition by actinomycin-D over the non-Hb areas and more important, an equal amount of fluctuation over the Hb region. Wilt (1967) has shown that there are predominantly 3 globin subunits in these primitive erythroblasts. The SDS gels reproducibly resolve these as two shoulders to the main subunit. Across these bands the ratio of $^{14}C/^{3}H$-leucine falls, indicating a differential inhibition by actinomycin-D of different globin subunits. Besides its implications with respect to the normal coordination of the synthesis of the Hb subunits these findings most probably reflect the requirement for continued synthesis of Hb mRNA during each of the seven generations of erythroblasts.

A second possible basis for the BrUdR-resistant program is the synthesis of a species of DNA by the hematocytoblast which does not replicate during subsequent erythroblast cell cycles. If made in excess, the proposed DNA might maintain a functional concentration despite the diluting effects of six divisions. This postulated species of DNA takes on the attributes of what are conventionally thought to be amplified genes, possibly for Hb itself. We have approached the problem in two ways. The first is by a direct search for gene amplification. The assay we have used is based upon the appearance of heavy-heavy DNA in cells treated for much less than a generation in BrUdR. Such a species, given semi-conservative replication, could be considered to be amplified DNA. As employed, the assay detects approximately 0.2% of the incorporated nuclear counts in HH-DNA. Unfortunately such a species does not obey all of the criteria required for an amplified gene. Thus, it has been found in all generations of developing erythroblasts, as well as in differentiating muscle cells, and terminally differentiated cartilage cells and fibroblasts. Its appearance is not correlated with periods of differentiation. Moreover, there is a 100-fold increase in HH-DNA after treatment of erythroblasts with FUdr for 20 h. It is difficult to say whether this is the "same" type of HH-DNA made in the absence of FUdr; however, only 2 h in FUdr results in a marked increase (3x) in its percentage. More recent data indicate that this DNase-sensitive species might be associated with the growing point of replication. We have detected a small amount of incorporation of Bu into the parental DNA strands at a level much higher than that of general repair. This probably represents the repair of nicks made on parental strands during the normal events of replication. Random breakage of the DNA during its isolation could then generate HH-DNA. This model would also account for the observed increase in HH-DNA after FUdr treatment.

A second way to approach the problem is to add $^{14}$C-TdR to eggs from 25 to 33 h of incubation (the period of BrUdR sensitivity). This is followed at 33 h by the addition of $^{3}$H-TdR. The embryo is then allowed to develop until the 3rd day when BrUdR is added for 24 h. The DNA is then isolated and run on a CsCl gradient and the ratio of $^{14}$C to $^{3}$H determined across the gradient. If there were a species of DNA that replicated only during the period of BrUdR sensitivity and was subsequently passed on to daughter cells, then this should appear as a high $^{14}$C/$^{3}$H ratio in the LL region of the gradient. This ratio should drop in the HL region and be lower still in the HH region. Figure 3 shows the results from a typical experiment of this kind. To the limit of resolution of 0.5% of the DNA (10,000 genes or 100 genes after a 100 fold divisional dilution) the $^{14}$C/$^{3}$H ratio is constant through the LL and HL areas of the gradient. The dip in the ratio over the HH regions represents the faster dividing cells in the 3rd day population. These are probably the last cells to undergo the hematocytoblast-erythroblast transition. Their low $^{14}$C/$^{3}$H ratio probably means that they were not dividing when only $^{14}$C/TdR was present. Although not conclusive, these experiments put an upper limit to the amount of DNA that could be responsible for a BrUdR-resistant program.

It is most likely that a third possibility is responsible for the "programmed events". We propose that BrUdR inhibits the appearance of a gene product whose synthesis is dependent on the appropriate quantal cell cycle and which is parceled out to the resultant daughter cells. Only the daughter cells of this particular quantal cell cycle are able to *initiate* the synthesis of Hb. Such a product is synthesized in the hematocytoblast during the period of BrUdR sensitivity and its continued synthesis is not required through the six subsequent generations of erythroblasts. Thus, assuming the substance to be a protein, there might be at least a 100-fold excess of this protein in early red cells as compared to the terminal ones. The appropriate double label experiments are presently being done in an effort to isolate such a gene

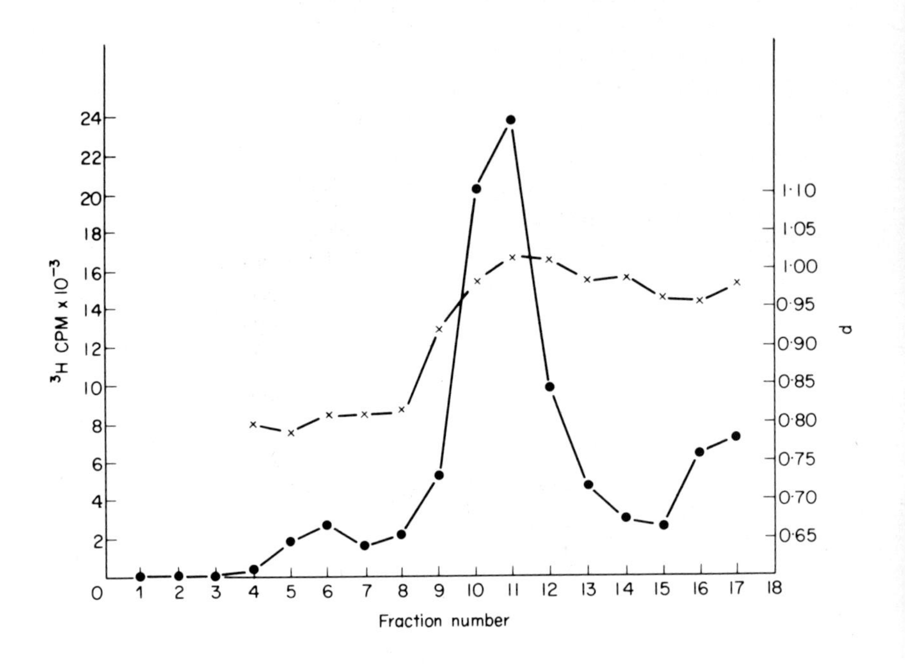


**Figure 3.** There is no detectable DNA that is synthesized early, but not synthesized later in the erythropoietic lineage. 5 $\mu$c of $^{14}$C-thymidine was injected into 25 h embryos. At 32 h, 50 $\mu$c of $^3$H-thymidine was also added. The embryos were allowed to develop until the third day, when 3 ml of BrUdR (2.5 mgm/ml) was added for one day. The nuclei cell were then isolated, washed, and the DNA extracted from the isolated nuclei by lysis in a 0.5% SDS, followed by RNase and pronase treatment and extraction in chloroform-isoanyl-alcohol. 4.5 ml of the dialyzed DNA solution in SSC was added to 6.85 gm of CsCl and run at 36,000 r.p.m. for 40 h in a 40 rotor and fractions were collected and counted as described in the Legend to Fig. 1. The solid line represents the $^3$H counts. The dashed line is the normalized ratio of $^{14}$C to $^3$H. The bulk of the radio-activity is found in the HL regions. About 5% is in the HH regions and about 10% is found complexed and remains on top of the gradient. LL DNA bands in fractions 14–16. The experiment has a resolution of 0.5% of the DNA. Unfortunately, higher doses of radioactivity sometimes lead to altered embryonic growth.

product. Should such a species exist and its mechanism of action be determined, only half of the BrUdR problem will have been solved for there will still remain the question of how BrUdR prevents the synthesis of this one species, leaving most of the remaining cell functions relatively intact.

## DISCUSSION

The results of experiments with presumptive myoblasts suggest that BrUdR compels cells to stay in proliferative cell cycles; that is, to produce more presumptive myoblasts (Holtzer, 1970b; Holtzer *et al.*, 1972a). Unless critical sites on the DNA are replaced by thymine, these cells are unable to enter the quantal cell cycle that yields myoblasts which synthesize myosin, actin and tropomyosin and which possess the capacity to fuse. BrUdR in the DNA of presumptive myoblasts keeps these cells in the penultimate generation of the myogenic lineage. According to this model, myogenic beta cells, retinal epithelial cells and hematocytoblasts with BrUdR in the respective critical sites of their DNA remain in their precursor stages and undergo proliferative cell cycles. As BrUdR was effective in isolating differences in the lineage of the red blood cells, so also it appears to distinguish particular points along these other two lineages. When BrU-containing presumptive myoblasts or BrU-containing chondroblasts (Holtzer, 1970b; Weintraub *et al.*, 1971) are reversed in normal medium, their progeny are capable of expressing the phenotypes of their respective lineages. We have presented data that indicate that such a protocol administered to myogenic-beta cells or retinal epithelial cells leads to cells that, though capable of replicating, do not necessarily undergo the next quantal cell cycle. Similar findings have been reported for primitive chondrogenic cells (Mayne *et al.*, 1972).

With respect to the amount of BrUdR remaining in the myogenic beta or retinal epithelial cultures after transfer to normal medium, it is possible to calculate that the average cell has only 3–4 BrU containing segregating units and that many cells should contain no BrU at all. Moreover, in those cells still containing BrU the probability that both homologues are substituted is extremely small. In order to explain the absence of phenotypic reversion by the residual amount of BrU remaining in the DNA, it is imperative to assume: (1) BrU in only one strand of the helix is an effective inhibitor. (2) Substitution in only one homologue of a given chromosome pair is sufficient for inhibition. (3) The determinants for the appearance of the given phenotype are located on many of the 70 odd chromosomes of these chick cells and moreover, these determinants are required in series.

The three systems examined in this paper describe how BrUdR can be used as a tool for studying the lineages of different cell types. Criteria such as "sensitivity to BrUdR" and "reversal of BrUdR inhibition" have been used to distinguish critical points in a given cellular history. Where BrUdR can be "reversed" it is reasonable to assume that the cellular potential is stable and such cells can be collected and used for extensive analysis in an attempt to distinguish them from their more easily isolated, terminally-differentiated progeny. Where BrUdR cannot be "reversed", two possibilities seem most likely: (1) There is a "cancellation" of the program for the given line. This loss of memory might be secondary to the degradative forces mediated over time or the dilutional effects of division. (2) Though replicating in normal medium, these cells lack in these tissue cultures the exogeneous cues required to allow the next quantal cell cycle. Such cues could lead to events as simple as the alignment of mitotic spindles for the generation of an asymmetric division or a directed migration of cells out of the generative epithelium. If the latter possibility is correct, then the potential of the cell is unaltered and these cells could again be compared to BrUdR-suppressed populations from later periods in the given lineage. If "cancellation" is due to dilutional effects of extra divisions, then it should be possible to mimic the effects of BrUdR by any means that add extra cell cycles.

Much of this discussion is highly speculative. And so it must remain until a better understanding is available of how BrUdR acts at a molecular level. Even without this information, however, this analogue is a curious tool to probe those historical processes that lead to a particular cell endowed not only with the machinery to translate for myosin, but with the machinery *not* to translate for norepinephrine or hemoglobin. To find cells with the capacity to produce these two latter luxury molecules, cells with different mitotic histories have to be examined.

## REFERENCES

Abbot, J. and Holtzer, H. (1968) *Proc. natn. Acad. Sci. U.S.A.* **59**, 1144.
Abbott, J., Mayne, R. and Holtzer, H. (1972). *Devl. Biol.* **28**, 430.
Bischoff, R. (1971). *Exp. cell Res.* **66**, 224.
Bischoff, R. and Holtzer, H. (1970). *J. cell Biol.* **44**, 134.
Campbell, G., Weintraub, H. and Holtzer, H. (1971). *J. cell Biol.* **50**, 669.
Coleman, J. and Coleman, A. (1970). *Exp. cell Res.* **59**, 319.
Hagopian, H. and Ingram, V. (1971). *J. cell Biol.* **51**, 440.
Hagopian, H., Lippke, J. and Ingram, V. (1972). *J. cell Biol.* **54**, 98.
Holtzer, H. (1970a). *In*: Cell Differentiation (eds O. Schjeide and J. De Villis) p. 476. Van Nostrand Reinhold Co., New York.
Holtzer, H. (1970b). *In*: Control Mechanisms in Tissue Cells (ed. H. Padykula). ISCB Symposium. Academic Press, New York and London.
Holtzer, H. and Sanger, J. (1972). *In*: Research in Muscle Development and Muscle Spindle (eds Banker and Przybylski). Excerpta Medica, Amsterdam.
Holtzer, H., Sanger, J., Ishikawa, H. and Strahs, K. (1972). *Cold Spring Harb. Symp. Quant. Biol.* (in press).
Lasher, R. and Cahn, R. (1969) *Devl Biol.* **19**, 415.
Mayne, R., Sanger, J. and Holtzer, H. (1971). *Devl Biol.* **25**, 20.
Mayne, R., Abbott, J. and Holtzer, H. (1972). *Exp. cell Res.* (in press).
Miura, Y. Wilt, F. (1971). *J. cell Biol.* **48**, 523.
Okazaki, K. and Holtzer, H. (1965). *J. Cyto. Histochem.* **13**, 726.

Shulte-Holthausen, H., Chacko, S., Davidson, E. and Holtzer, H. (1969). *Proc. natn. Acad. Sci. U.S.A.* **63**, 864.
Stockdate, F., Okazaki, K., Nameroff, M. and Holtzer, H. (1964). *Science, N.Y.* **146**, 533.
Weintraub, H., Campbell, G. and Holtzer, H. (1971). *J. Cell Biol.* **50**, 652.
Wilt, F. (1967). *Adv. Morph.* **6**, 89.
Wenk, M. (1971). *J. cell Biol.* **49**, 673.

FEBS Symposium, Volume 24, 1972, pp. 55-76

# On the Mechanism of Biosynthesis and Transport of Messenger RNA in Eukaryotes

G. P. GEORGIEV, E. M. LUKANIDIN and A. P. RYSKOV

*Institute of Molecular Biology,*
*Academy of Science U.S.S.R., Moscow U.S.S.R.*

## I. INTRODUCTION. THE GENERAL SCHEME OF GENE EXPRESSION IN EUKARYOTES

This report is devoted to new experimental data concerning structure of the transcriptional units in mammalian cells and the mechanism of mRNA transport. The experiments were developed to check some predictions following on from the theoretical models recently described. We start with a short description of these models. The first hypothesis concerns the organization of the transcriptional unit (Georgiev, 1969).

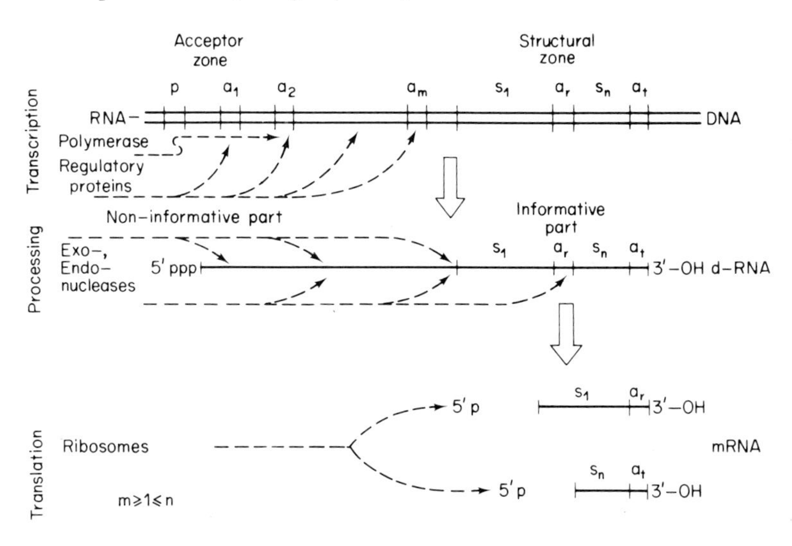


**Figure 1.** Slightly modified model of the structure of a transcriptional unit in eukaryotes. p: promoter; a, $a_1$, $a_2$, $a_m$: acceptor sites; $a_r$: site recognized by specific endonucleases; $a_t$: terminator site; $S_1$, $S_n$: structural cistrons.

It is postulated that transcriptional units in eukaryotes have a complex structure. They consist of sequences of two kinds: acceptor sites, mainly located near the promoter (acceptor, or non-informative zone), and structural cistrones, localized distally of the promoter (structural or informative zone). Many of the acceptor sites correspond to operators with which regulatory

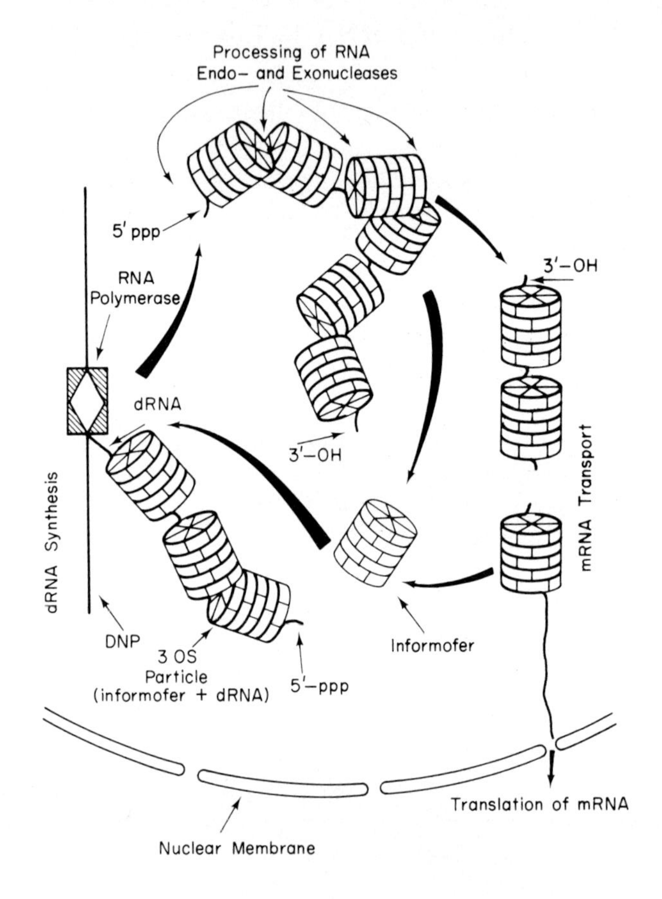


**Figure 2.** Scheme of the mRNA transport in mammalian cells.

proteins (or regulatory polynucleotides) may interact. Non-informative loci may be represented also by some other "service" sequences such as terminators, sequences recognized by specific ribonucleases or by protectors, etc. In this case they may be interspaced by structural genes. If the RNA-polymerase does not meet obstacles in its way, the transcription of the whole operon, including the non-informative part, takes place. A large dRNA precursor is formed, which is a copy of both the informative (mainly in the 3′-side) and the non-informative (mainly in the 5′-side) zones. Then most of the non-informative part of the dRNA is degraded ($dRNA_2$) and the informative part is transferred to the cytoplasm ($dRNA_1$ or true mRNA) (Fig. 1).

The transcription of the structural genes is under the control of the interaction between regulatory proteins and acceptor sites corresponding to operators. Only after the RNA polymerase has moved along the whole acceptor zone, may the structural genes be also transcribed.

It was also postulated that some acceptor sites correspond to the repetitive DNA base sequences, namely to their so called "intermediate" or "kinetic" fraction. The occurrence of the same sequence in many different transcriptional units may provide a mechanism for the massive switching on or off of genes during cell differentiation.

The second model concerns the mechanism of mRNA transport (Georgiev and Samarina, 1971). Following its synthesis, the dRNA chain is combined with the informofers, which are specific macroglobular protein particles. One informofer is combined with one dRNA chain of about $2 \times 10^5$ dalton. The dRNA distributed on the surface of the informofer is attacked by the processing enzymes: endo- and exo-nucleases. This leads to the breakdown of the nucleus-restricted non-informative dRNA. The true mRNA complexed with the informofer is transferred first to the nuclear membrane and then to the cytoplasm. Informofers probably do not leave the cell nucleus (Fig. 2).

Many earlier data are in agreement with the above models. In this report we present the results of new experiments, specifically developed to check some predictions, which follow from these models.

## II. THE DATA ON THE STRUCTURE OF THE TRANSCRIPTIONAL UNIT IN MAMMALIAN CELLS

From the model described above (Fig. 1) it follows that (1) the 5′-end part of the newly formed nuclear dRNA should be enriched in replicas from reiterated DNA base sequences, which are not transferred to the cytoplasmic polysomes; (2) the 3′-end part should contain sequences which correspond to the true mRNA and therefore are transferred to the cytoplasm.

To check these predictions a new approach has been developed consisting of the hybridization analysis of the end-labeled nuclear dRNA. The 5′-end may be detected in $^{32}P$ labeled RNA as it contains triphosphate groups (Ryskov and Georgiev, 1970); the 3′-end may be easily labeled with $^3H$ by means of periodate oxydation-$^3H$ bohrhydride reduction technique (Leppla *et al.*, 1968).

The experiments have been done on the newly-formed dRNA which represents copies of the operons. For its isolation a slightly modified hot phenol fractionation procedure was used (Georgiev and Mantieva, 1962; Arion *et*

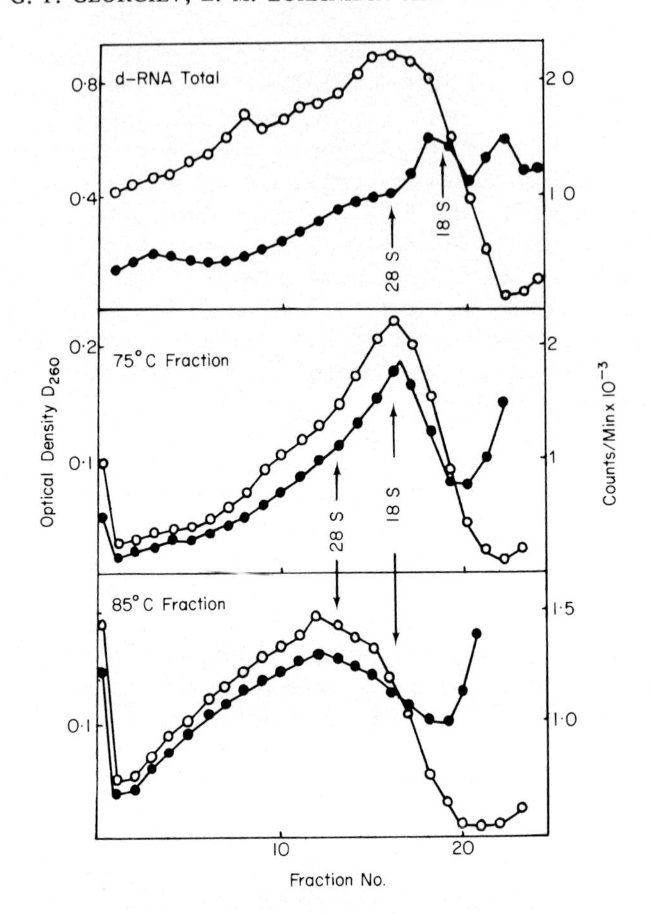


**Figure 3.** Sedimentation profiles of dRNAs isolated from rat liver cells A: Total nuclear dRNA (isolated at 85°C) labeled with $Na_2H^{32}PO_4$ for 2.5 h. B and C: 75°C fraction (interval 55-75°C) and 85°C fraction (interval 75-85°C), labeled with $Na_2H^{32}PO_4$ for 36 h. Centrifugation in Spinco-L2, rotor SW 25.2, 16,000 rpm, 22°C for 12 h (A) or 14 h (B, C) 5-25% sucrose in 0.25% SDS, 0.05 M NaCl, 0.0015 M EDTA, 0.01 M tris, pH 7.5. ●—● optical density at 260 nm (in the top fractions it hydrolyzed DNA if present; ○—○ radioactivity ($^{32}P$).

RNA fractions were isolated by the hot phenol fractionation procedure (Georgiev and Mantieva, 1962). The suspension of Ehrlich carcinoma cells or the rat liver homogenate in 0.14 M NaCl was treated with phenol at 4°C and the interphase obtained (phenolic nuclei) treated with 0.14 M NaCl-phenol, pH 6, twice at 4°C, once at 40°C and once at 55°C. The "phenolic nuclei" were then suspended in 1% SDS 0.14 M NaCl and shaken with an equal volume of phenol at 65°C or at 75°C for 15 min.

The water phase was collected and the interphase again treated in the same way but at 85°C. In some cases the interphase was treated directly at 85°C (without pretreatment at 65°C or 75°C). The yield of nuclear RNA was 90-100%. The fractions obtained were deproteinized by additional treatments with phenol and chloroform, precipitated by ethanol, treated with DNase (Whorthington, RNase-free), and ultracentrifuged in a *al.*, 1967). In Fig. 3 the sedimentation profiles of dRNA fractions isolated from rat liver and Ehrlich ascited carcinoma cells, obtained after rather short and long-term labeling, are presented. It can be seen that dRNA may be isolated by this procedure in a rather non-degraded state. The peak of distribution of the material in the 85°C-fraction lies in the 30-35S zone and a significant part of it is found also in the 40-70S zone. These figures are similar to those obtained for dRNA by other authors who used milder methods of isolation, which however did not allow them to isolate dRNA non-contaminated by rRNA. Thus the hot phenol treatment itself even at 85°C produces only a slight degradation of RNA. In the 65°C (75°C) fractions the peak of distribution is localized in the 18S zone, although part of the material sediments in the 40-70S zone.

Since the base composition of the RNA obtained is similar to that of DNA (G + C/A + U = 0.8), one can conclude that the hot phenol fractionation allows one to isolate giant (molecular weight $2 \times 10^6$ dalton) as well as rather low molecular weight ($0.4\text{-}0.8 \times 10^6$ dalton) fractions of nuclear dRNA.

**Only giant dRNA is the primary product of dRNA synthesis**

It is known from experiments in cell-free systems and also on viral and some cellular RNAs, that the primary product of transcription contains a triphosphate group at the 5′-end (Maitra and Hurwitz, 1965; Takanami, 1966; Hatlen *et al.*, 1969). Under alkaline hydrolysis nucleoside tetraphosphate is formed (pppXp). In the mature molecule the $\beta$- and $\gamma$-phosphates are usually absent and the alkaline hydrolysis results in the formation of nucleoside-di-(pXp) or monophosphates (Xp). All these compounds may be separated by chromatography on DEAE-Sephadex in urea (Roblin, 1968).

We have found that alkaline hydrolysates of newly formed dRNA from rat liver (after 2-3 h labeling) contain both the pppXp and pXp. Then the total nuclear dRNA was separated by density gradient ultracentrifugation into three zones: heavy (>30S), intermediate (20-30S), and light (10-18S) and each of them was studied. Only hydrolysates obtained from the heavy fraction contain pppXp, while the content of pXp in this fraction is rather low. On the other hand hydrolysates of the other two fractions contain more pXp but do not contain pppXp at all (Table 1).

Thus the true starting points of transcription may be found only among heavy dRNAs. In other words, it seems likely that only heavy dRNAs with

---

SDS-sucrose gradient. The fractions from the density gradient were collected, analysed, precipitated and used for further experiments. The time from the beginning of the experiment until ultracentrifugation was one day. If RNA samples were stored for a longer time they were pretreated with bentonite.

**Table 1.** The distribution of $^{32}P$ radioactivity between nucleoside mono-, di-, and tetraphosphates in the alkaline hydrolysates of nuclear d-RNA fractions.

| | Radioactivity | | | | |
|---|---|---|---|---|---|
| | Xp | pXp | | pppXp + ppXp | |
| RNA Fractions | cpm | cpm | per cent of total | cpm | per cent of total |
| Heavy dRNA | 1,440,000 | 187 | 0.013 | 590 | 0.041 |
| Light dRNA | 600,000 | 600 | 0.1 | 25 | 0.004 |

dRNA was isolated from rat livers 3 h after injection of $Na_2H^{32}PO_4$ (1 mCi per animal) by the hot phenol fractionation technique (Georgiev and Mantieva, 1962; Arion *et al.*, 1967). After ultracentrifugation the heavy (35S) and light (30S) fractions were collected, purified by Sephadex G-75 gel filtration, precipitated by 5% trichloroacetic acid (TCA) and hydrolysed in 0.5 N KOH for 24 h at 37°C. A mixture of non-labeled oligonucleotides was added (RNA digested by pancreatic RNase) to the neutralized hydrolysates and the material separated on a DEAE-Sephadex column in 7 M urea (Roblin, 1968). Elution with a salt gradient gave a number of peaks with an increasing number of nucleotides in the chain.

The material of peaks III (trinucleotides) and IV to VI (tetra to hexanucleotides) was collected, desalted, and re-hydrolysed with 0.5 N KOH under the same conditions to destroy all surviving labeled oligonucleotides. The chromatography was repeated. At this step of purification peak III contained, exclusively, labeled nucleoside diphosphates (pXp) and peaks IV and V, labeled nucleoside tetraphosphates (pppXp) or triphosphates (ppXp).

molecular weight $>2.0 \times 10^6$ are the primary products of transcription, and the light chains are formed as a result of processing of giant dRNAs.

The hybridizability of the 5′-ends of the newly formed giant dRNA was determined and compared with the hybridizability of the internal sequences of dRNA.

It is well known that under the usual conditions of hybridization (with $C_ot$ values of about 10-100) only those RNA sequences are hybridized which are synthesized on the repetitive DNA sequences (Britten and Kohne, 1968; Ananieva *et al.*, 1968; Church and McCarthy, 1968). Thus the hybridizability in the presence of DNA excess is proportional to the content of RNA synthesized on these sequences. It was found that the alkaline hydrolysates of RNA hybridized in the presence of excess of DNA contain four to ten times more pppXp groups per Xp group than alkaline hydrolysate of non-hybridized RNA (Table 2). Thus 5′-end sequences are concentrated in the hybridized material. In some recent experiments in which higher $C_ot$ values (100-200) were used, the hybridization of end sequences containing pppXp groups reached 50% or even more. This result strongly suggests that all or most of

**Table 2.** Hybridizability of 5′-end sequences in heavy nuclear dRNA from rat liver

| Experiment number | RNA fraction | Xp cpm × $10^{-6}$ | pppXp + ppXp cpm | pppXp + ppXp per cent of total activity | pXp cpm | pXp per cent of total activity | Hybridization per cent: total RNA | Hybridization per cent: pppXp |
|---|---|---|---|---|---|---|---|---|
| 1 | Hybridized | 0.11 | 130 | 0.118 | 30 | 0.027 | 11 | 35 |
| | Nonhybridized* | 0.52 | 140 | 0.027 | 110 | 0.021 | – | – |
| 2 | Hybridized | 0.17 | 108 | 0.062 | 61 | 0.035 | 7 | 20.5 |
| | Nonhybridized* | 0.39 | 68 | 0.018 | 272 | 0.07 | – | – |
| 3 | Hybridized | 0.057 | 154 | 0.27 | – | – | 7.4 | 48.5 |
| | Nonhybridized | 0.69 | 160 | 0.023 | – | – | – | – |

In experiments 1 and 2 only, part of the non-hybridized material was analysed. The conditions of hybridization: experiment 1–170 mg DNA and 3.4 mg dRNA annealed in 50 ml 2 x SSC for 14 h; experiment 2–100 mg DNA and 2.0 mg dRNA annealed in 30 ml 2 x SSC for 14 h; experiment 3–100 mg DNA and 2.0 mg dRNA annealed in 15 ml 7 M urea, 4 x SSC, 0.1% SDS for 96 h at 41°C (according to a technique by Kurilsky and Gros). In all cases DNA-gels cross-linked by UV irradiation were used. The loss of $\gamma$ P from ATP in the conditions mentioned above does not exceed 10-15%.

the transcriptional units begin with reiterated DNA base sequences and thus supports one of the predictions of the model. A similar conclusion has been achieved in our laboratory by means of the analysis of non-completed dRNA molecules synthesized after UV irradiation of Ehrlich carcinoma cells (Mantieva *et al.*, 1969).

The pXp content in the hybridized and non-hybridized RNA is almost the same. Thus, at least a significant part of pXp groups does not originate from a dephosphorylation at the starting sequences of dRNA but rather from the internal parts of the precursor molecules after a nuclease attack.

A most important question is whether or not the starting sequences in dRNA are transferred to the cytoplasm. The preliminary results of competition experiments indicate that the polysomal RNA does not decrease the binding of sequences containing pppXp. Thus it seems very probable that they do not carry any structural information for protein synthesis, which is in agreement with the model discussed.

**The 3′-end of the giant dRNA is predominantly transferred to the cytoplasm**

To analyse the fate of the end part of the dRNA molecule, the giant dRNA has been labeled in 3′-end position by the periodate oxydation–$NaB^3H_4$ reduction technique. Such double labeled RNA ($^{14}C$ or $^{32}P$ randomly, $^3H$ in the 3′-end) was used in hybridization-competition experiments.

**Table 3.** Base composition of 3′-ends in nuclear dRNA

| dRNA source | Fractions of nuclear dRNA | 3′-end nucleoside per cent A | U | G + C |
|---|---|---|---|---|
| Rat liver | Total | 70 | 20 | 4 + 6 |
| | Heavy (⩾ 35S) | 71 | 13 | 16 |
| | Intermediate (20-30S) | 77 | 14 | 9 |
| | Light (12-20S) | 68 | 19 | 13 |
| Ehrlich carcinoma cells | Heavy, hybridized | 65 | 19 | 16* |
| | Heavy, non-hybridized | 62 | 15 | 23* |
| | Light, hybridized | 75 | 17 | 8 |
| | Light, non-hybridized | 80 | 12 | 8 |

RNAs were labeled with $^3H$ at the 3′-end position; 3′-end nucleosides were removed from nucleotides on Dowex-1 (formiate) column and chromatographed on Whatman 3 MM in two systems: (1) tert. butanol, methylethylketone, $H_2O$, conc. HCOOH (44:44:15:0,26); (2) tert. butanol, methylethylketone, $H_2O$ conc. $NH_4OH$ (40:30:20:10). The spots were excised and counted in dioxane scintillator.

* These figures are slightly overestimated as a result of radioactive contaminations.

**Table 4.** The results of typical experiments on competitive hybridization of 3′-ends in nuclear dRNA

| Material | dRNA fraction | Conditions of hybridization | | | | Hybridized RNA cpm | | | Non-hybridized RNA cpm | | | Per cent of radioactivity in hybrid. | | Per cent of competition | |
|---|---|---|---|---|---|---|---|---|---|---|---|---|---|---|---|
| | | DNA mg | dRNA mg | Cyt. RNA mg | Vol. ml | $^{14}C$ | $^{3}H$ | $^{14}C/^{3}H$ | $^{14}C$ | $^{3}H$ | $^{14}C/^{3}H$ | $^{14}C$ | $^{3}H$ | $^{14}C$ | $^{3}H$ |
| Ehrlich | Heavy | 13 | 0.25 | – | 2.7 | 214 | 494 | 0.44 | 3,940 | 2,200 | 1.8 | 5.2 | 18.3 | – | – |
| ascites | (⩾ 35S) | 13 | 0.25 | 18 | 2.7 | 194 | 155 | 1.25 | – | – | – | 4.7 | 5.7 | −10 | −69 |
| carcinoma | Light | 10 | 0.20 | – | 2.2 | 74 | 940 | 0.08 | 2,210 | 3,940 | 0.56 | 3.2 | 19.2 | – | – |
| cells | (18S) | 10 | 0.20 | 15 | 2.2 | 58 | 430 | 0.14 | – | – | – | 2.5 | 8.8 | −22 | −54 |
| Rat | Heavy | 8 | 0.16 | – | 2.7 | 6,500 | 310 | 21.0 | 45,000 | 863 | 52.0 | 12.7 | 26.5 | – | – |
| liver | (⩾ 35S) | 8 | 0.16 | 8 | 2.7 | 6,600 | 202 | 32.6 | – | – | – | 12.8 | 17.2 | 0 | −35 |
| | Light | 5 | 0.09 | – | 2.2 | 1,700 | 606 | 2.8 | 13,400 | 2,045 | 6.6 | 11.3 | 22.9 | – | – |
| | (18S) | 5 | 0.09 | 5 | 2.2 | 1,100 | 352 | 3.1 | – | – | – | 7.2 | 13.3 | −37 | −42 |

The fractions of dRNA isolated after 40 h incubation with $^{14}C$-orotic acid were labeled at the 3′-end position with $^{3}H$ (Leppla *et al.*, 1968). They were extensively purified (additional DNAse-pronase treatment; phenol and chloroform deproteinization; 2.5 M NaCl reprecipitation; gel-filtration through Biogel P-100 column; reprecipitation with 5% TCA in the cold, rapid removal of TCA by washing with Ethanol-0.2 M Na acetate), and hybridized with homologous DNA with or without non-labeled cytoplasmic polysomal RNA used as competitor. DNA-gels cross-linked by UV-radiation were used (Arion *et al.*, 1967). After hybridization all samples were hydrolysed in 0.5 N KOH and neutralized with $HClO_4$; $^{3}H$-labeled nucleosides were purified by ion-exchange and paper chromatography and counted. All $^{3}H$-counts represent the sum of purified A and U derivatives.

It was found that after alkaline hydrolysis of the tritiated RNA about 70-75% of the label in the hybrid as well as in the non-hybridized material belongs to adenosine and about 20% to uridine derivatives (Table 3). Thus, different dRNA molecules in rat liver as well as in Ehrlich carcinoma cells have the same end nucleotides. Both giant and light dRNA have the same end nucleotides. This suggests that the 3′-end is conserved during processing.

The hybridizability of the $^{3}$H-labeled 3′-end is higher than that labeled with $^{32}$P or $^{14}$C. (Table 4). This indicates that the 3′-ends are enriched in copies from repetitive DNA base sequences and different dRNAs may have the same or very similar end sequences

On the other hand the competition of the polysomal RNA with the 3′-end sequences of giant dRNA is much stronger than with total dRNA. This means that the true mRNA ($dRNA_1$) is predominantly localized near the 3′-end of nuclear dRNA, which again is in line with our postulated model.

The data presented support the following conclusions: (1) dRNA is synthesized only in the form of giant precursor molecules, the shorter chains being the result of processing (data on the nature of 5′-ends). (2) A significant part of $dRNA_1$ or true mRNA is localized at the 3′-part of dRNA precursor (results of competition experiments with 3′-ends). (3) 5′-ends of newly formed dRNA are enriched in copies of repetitive DNA base sequences (data on the high hybridizability of the 5′-ends). (4) The copies of the starting sequences of the operons probably do not reach the polysomes and do not contain mRNA; thus they are non-informative (the results of preliminary experiments on the absence of competition between polysomal RNA and triphosphorylated 5′-ends of giant nuclear dRNA).

These findings, although not proving our model as a whole, do support some of the predictions following from it.

The results obtained in the present work indicate that repetitive base sequences are also present at the ends of several transcriptional units; indeed the hybridizability of the 3′-ends of newly formed giant dRNA is higher than that of total dRNA. It may be that some operon ends contain "service sequences" responsible for the termination of RNA synthesis (terminators) or for the RNA protection during processing. These sequences should be rather short, as the strong inhibition of their hybridization by cytoplasmic RNA is not associated with any significant decrease of total RNA bound to DNA. The question requires further investigation. The presence of relatively short sequences at the ends of the operons, whose copies are transferred to the cytoplasm together with true mRNA, may explain the fact that the cytoplasmic mRNA is hybridized to some extent with DNA (Williamson and Morrison, 1970), which could not be explained on the basis of the original model.

## III. ON THE MECHANISM OF mRNA TRANSPORT

**The isolation of informofers free from dRNA and their properties**

It was shown earlier that almost all nuclear dRNA is combined with informofers (Samarina *et al.*, 1968) (Fig. 2). However, the informofers could not be isolated in a pure state free from dRNA. The removal of RNA by RNase led to the aggregation of protein particles.

It has now become possible to isolate the informofers free from RNA by means of treatment with 2 M NaCl. It is known that the protein of informofers (called informatin) is labeled very slowly (Samarina *et al.*, 1965). In order to label it, Ehrlich ascites carcinoma cells were exposed to $^{14}C$-protein hydrolysate for 2 days. In another experiment 30S particles isolated from rat liver containing orotate-$^{14}C$-labeled RNA were labeled *in vitro* with $^{125}I$ according to Bale *et al.* (1966).

The 30S particles were further treated with 2 M NaCl and ultracentrifuged through a sucrose density gradient containing 2 M NaCl (Fig. 4). All RNA label was recovered in a peak with sedimentation coefficient of about 6-10 S with a tail in the heavy zone. On the other hand the main part of the labeled protein sedimented as a homogeneous peak with sedimentation coefficient of about 30S. In CsCl density gradient (Fig. 5) all labeled RNA was recovered near the bottom whereas the labeled protein again banded as a sharp peak with a buoyant density of about 1.34 g/cm$^3$. Thus at high salt concentration the full dissociation of dRNA does not result in the destruction or aggregation of informofers and in fact the latter may be obtained in a pure state as high molecular weight protein particles.

According to electron microscopic observations (Fig. 6) free informofers do not differ from the complete 30S particles. They appear as homogeneous particles of about 180 Å in diameter. It was also found that free informofers are able to interact with free exogeneous dRNA by removing the excess of NaCl by dialysis against 0.1 M NaCl – 0.001 M $MgCl_2$ in 0.01 tris pH 7.5 (Fig. 7).

After the addition of urea the informofers are dissociated into relatively low molecular weight (MW ~ 40,000) subunits. They are identical with informatin as shown by electrophoresis in polyacrylamide gel (Fig. 8).

The isolation of free informofers gives strong support to the proposed structure of nuclear particles containing dRNA (Fig. 2). It is clear that the removal of all RNA without destruction of particles could not have been obtained if the RNA were not localized at the surface of the protein globule. The localization of dRNA at the surface of the informofers explains how it may be easily attacked by nucleolytic enzymes participating in dRNA processing.

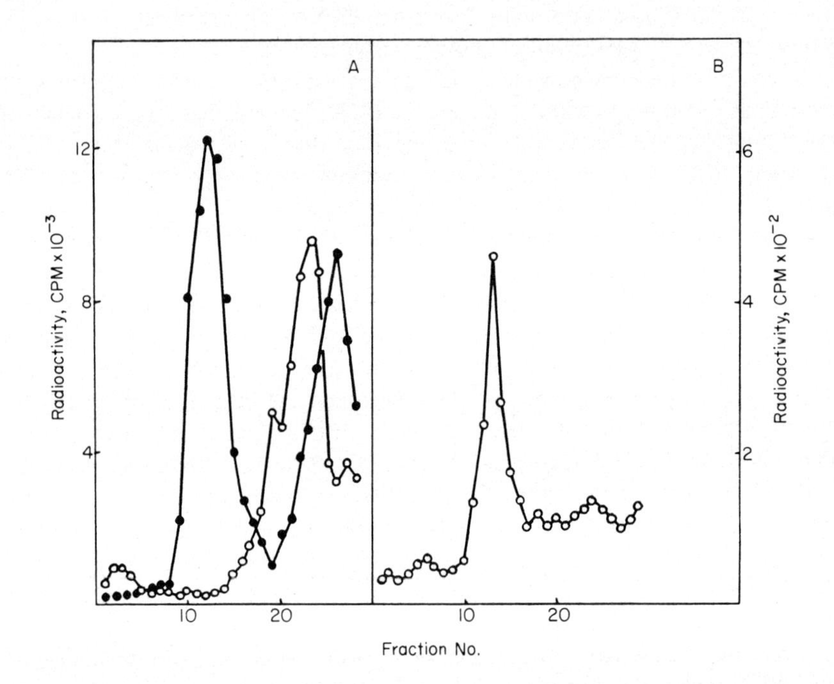


**Figure 4.** Sedimentation distribution of free informofers. Ribonucleoprotein 30S particles were dialysed against 2 M NaCl; the mixture was layered on a 15-30% sucrose gradient in 2 M NaCl and ultracentrifuged for 14 h at 24,000 rpm in SW 25-rotor of Spinco L-2 ultracentrifuge at 2°C. A: Double-labeled particles from rat liver containing $^{14}$C-RNA and $^{125}$J-protein (compare with Fig. 1). ●—● $^{125}$J, cpm ○—○ $^{14}$C, cpm. B: Ehrlich carcinoma particles containing $^{14}$C-labeled protein. ○—○ $^{14}$C, cpm.

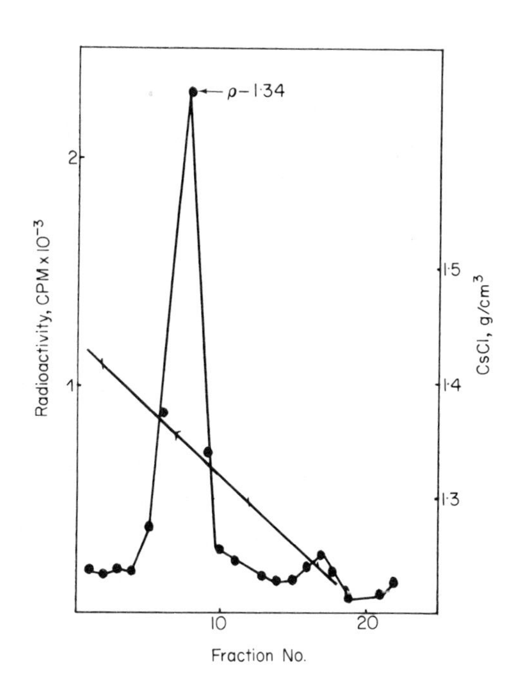


**Figure 5.** Banding of free informofers in CsCl density gradient. The peak from the sucrose gradient (Fig. 2A) was collected and fixed with 2% $CH_2O$ (the same result may be obtained without fixation); 1 ml was layered on 4 ml preformed CsCl density gradient ($\rho$ – 1.4 – 1.2 g/cm$^3$), and centrifuged in the SW-50-rotor of Spinco L-2 ultracentrifuge for 18 h at 45,000 rpm at 2°C. ●—● $^{125}$J, cpm.

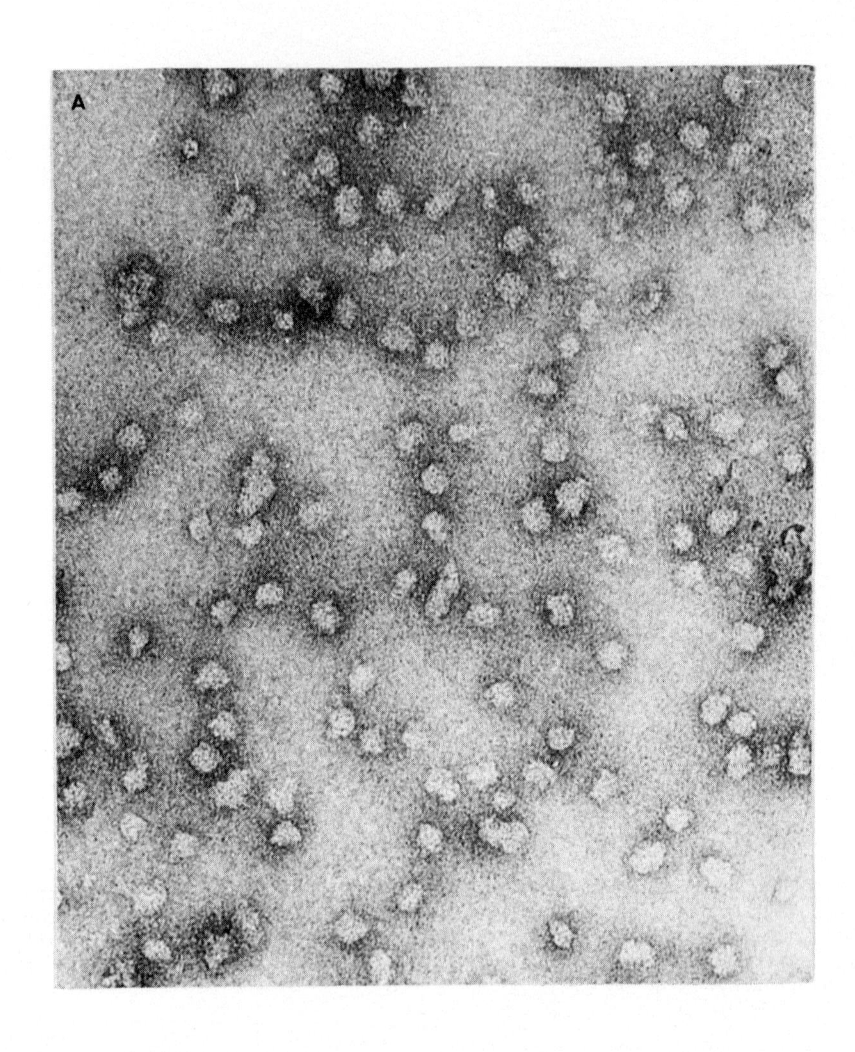


**Figure 6.** Electron micrography of (A) informofers and (B) 30S nuclear particles

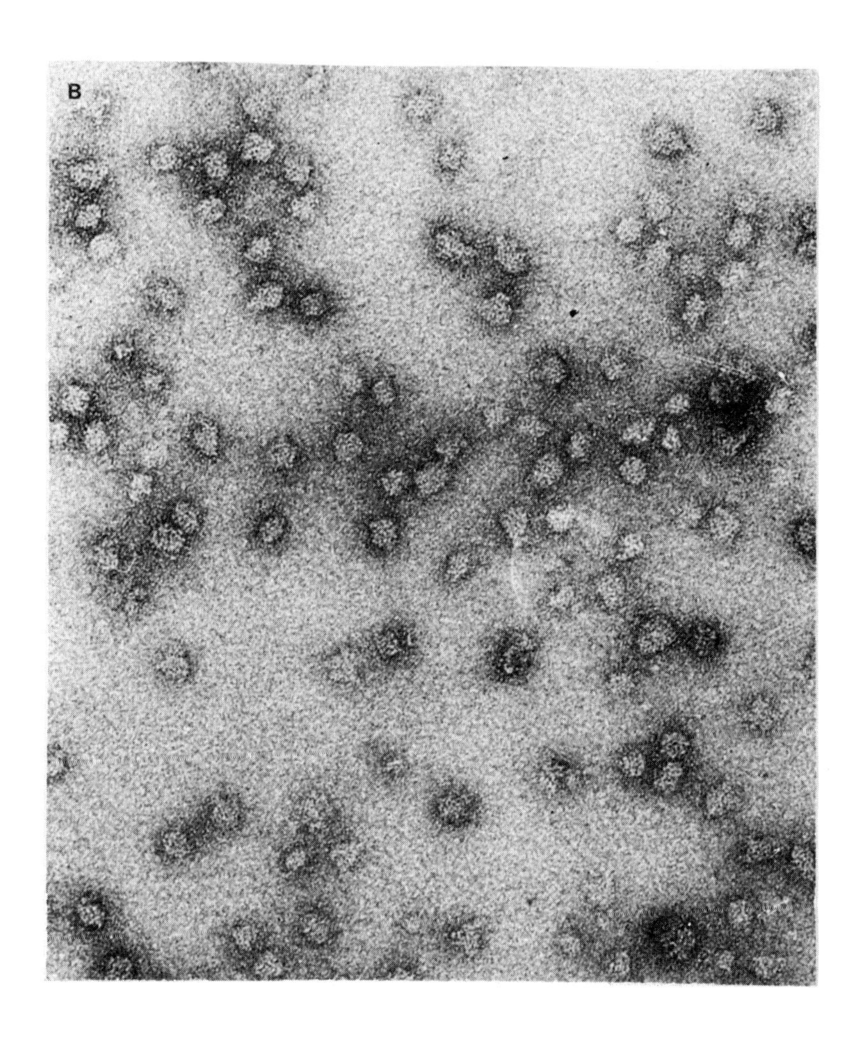


obtained from sucrose gradient. Negative stained preparation (Magnification: x 200,000).

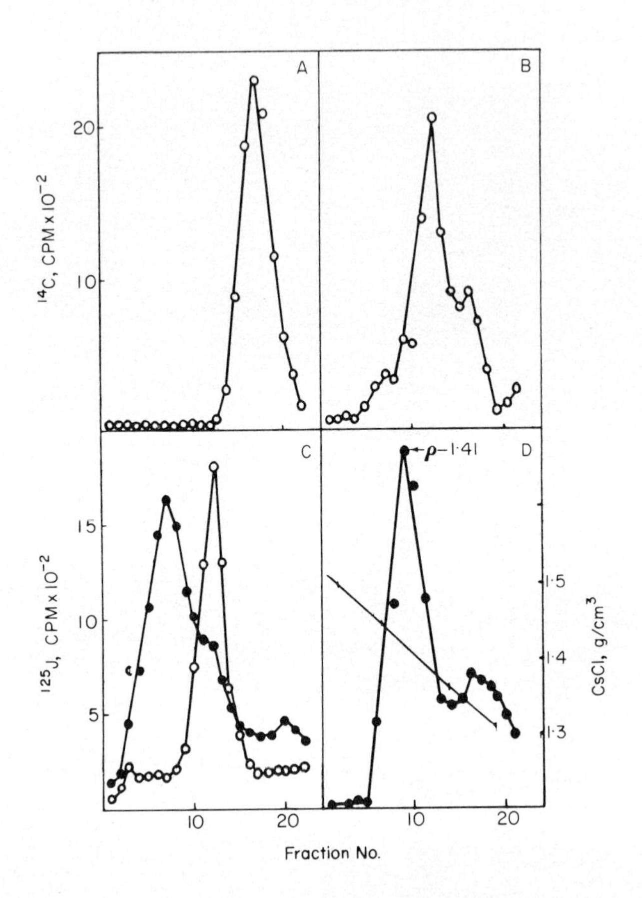


**Figure 7.** Sedimentation and density distribution of the products formed as a result of mixing free informofers and dRNA. Free rat liver informofers were mixed in 2 M NaCl with purified dRNA and dialysed against 0.1 M NaCl, 0.001 M $MgCl_2$, 0.025 M phosphate pH 7.5. A: Sedimentation distribution of $^{14}$C-dRNA used in reconstitution experiment (rotor SW-50, 15-30% sucrose gradient, 50,000 rpm for 2 h at 2°C). B: Sedimentation distribution of material obtained after mixing $^{14}$C-dRNA from (A) with non-labeled informofers [conditions are the same as in (A)]. C: The same as (B) but after mixing non-labeled dRNA with $^{125}$J-labeled informofers. o——o 6-10S dRNA (isolated from 30S particles) was used for reconstitution; •——• 12-18S dRNA obtained by the hot phenol method was used. D: Centrifugation through a CsCl density gradient (1.3-1.5 g/cm$^3$) (SW-50 for 18 h, 45,000 rpm at 2°C) of material reconstituted from informofers and 12-18S dRNA.

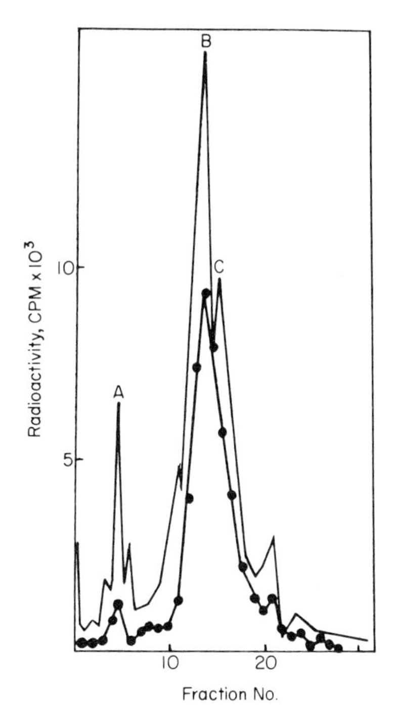


**Figure 8.** Polyacrylamide gel electrophoresis of labeled protein from purified informofers. The proteins of free informofers labeled with $^{125}J$ were obtained by urea treatment, were incubated with mercaptoethanol and mixed with carrier proteins from 30S particles not treated with mercaptoethanol.—Optical density (peaks A, B and C are visible); ●—● Radioactivity (The bulk of material is concentrated in peak B).

**Adenovirus-specific dRNA is complexed with informofers inside the cell nucleus.**

It was shown in competition–hybridization experiments that both types of nuclear dRNA–dRNA$_1$ (precursor of mRNA) and dRNA$_2$ (the nucleus-restricted RNA)–are complexed with informofers (Mantieva *et al.*, 1969; Dreus, 1969).

However, hybridization experiments reveal only copies from repetitive DNA sequences. It seems thus interesting to look at the product of certain genes. One available model is offered by the cells infected with a DNA virus such as the Adenovirus localized inside the cell nucleus. It was found that the virus-specific RNA is synthesized in the nucleus of infected cells and that all the sequences present in this RNA reach the polysomes (Parsons *et al.*, 1971).

Thus this RNA may be considered as a true mRNA. To study the nature of nuclear complexes containing dRNA (Zalmanzon *et al.*, 1972) the Adenovirus infected cells were labeled with uridine-$^3$H and the cell nuclei isolated. To extract the dRNP complexes the nuclei were treated three times with a 0.1 M NaCl solution, containing 0.001 M $MgCl_2$ in 0.01 M tris (pH 8.0) and 2% Triton X-100 at 35°C for five minutes. Under these conditions the main part of nuclear dRNP is solubilized. Extraction without Triton X-100 or in the cold does not release dRNP from cultured cells in contrast to rat liver, Ehrlich carcinoma cells etc.

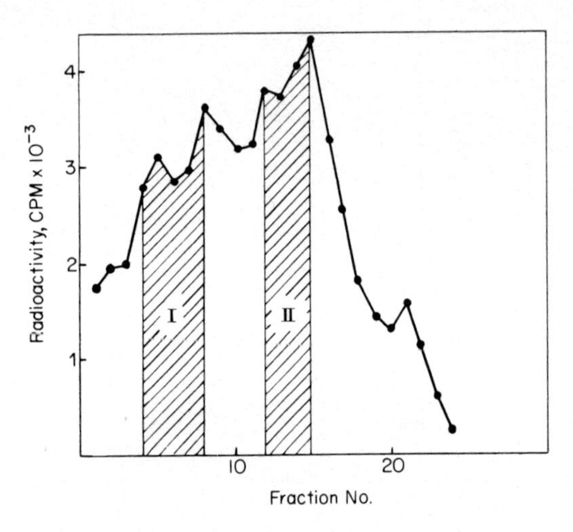


**Figure 9.** Sedimentation profile of nuclear extracts of Adenovirus infected FL cells. The extract was layered on top of 15%-30% sucrose-gradient containing 0.1 M NaCl, 0.001 M $MgCl_2$ and 0.01 M triethanolamine (pH 7.8). The gradients were centrifuged in the SW-50 rotor at 45,000 r.p.m. for 2 h. Fractions I and II were collected for the centrifugation in CsCl gradient.

The extracts were ultracentrifuged through a sucrose density gradient. The labeled material was recovered in 30S particles, as well as in heavier structures, representing polyparticles or complexes of dRNA with a number of informofers (Fig. 9). The survival of poly-particles even in the absence of RNase inhibitor probably depends on the low RNase activity in the cells used. All the particles (30S and 40-100S) after fixation with formaldehyde are banded in CsCl density gradient as a single homogeneous peak with a buoyant density of 1.41-1.42 g/cm$^3$ (Fig. 10). The same is the buoyant density of nuclear complexes from rat liver after treatment with Triton X-100 (Olsnes, 1970).

According to several other tests (electron microscopy RNase sensitivity, properties of the proteins) the particles from FL cells could not be distinguished from the complexes of dRNA with informofers obtained from rat liver.

The RNA was isolated from different fractions of sucrose and CsCl density

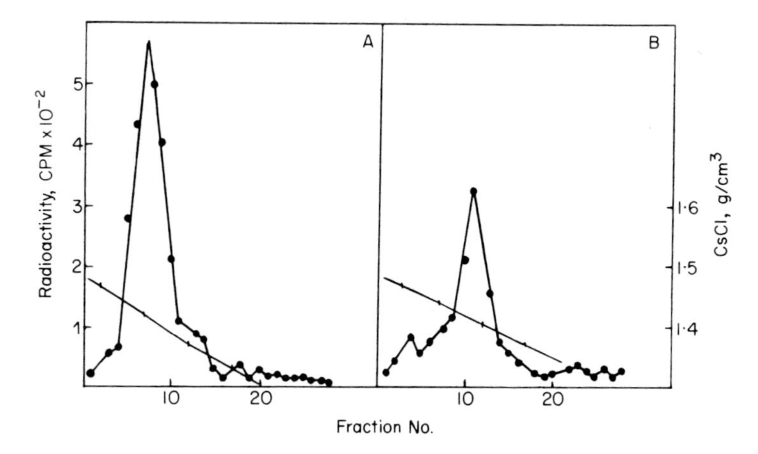


**Figure 10.** CsCl density gradient centrifugation of nuclear RNP particles from infected cells. RNP particles were isolated as in Fig. I, fixed with formaldehyde, layered on preformed CsCl density gradient (1.3-1.6 g/cm$^3$) and centrifuged at 45,000 r.p.m. for 16 h. A: Fraction II and B: Fraction I from the sucrose gradient.

gradients. The later became possible by decreasing to 15-20 h the length of the treatment with formaldehyde which was followed by ethanol precipitation of the particles and treatment with pronase and SDS-phenol.

The RNAs were hybridized with the host or virus DNA. As is shown in Table 5, virus-specific RNA was found in complexes with informofers: namely

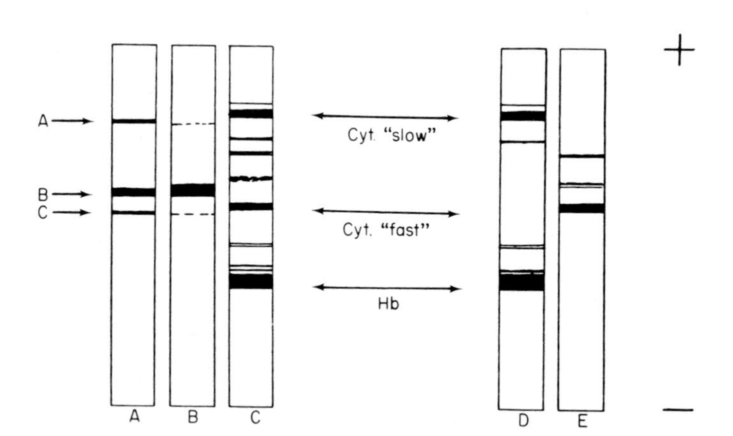


**Figure 11.** Electrophoresis of nuclear and cytoplasmic proteins in polyacrylamide gel. A, B: Proteins of 30S particles from rat liver before (A) and after (B) mercaptoethanol treatment. C: Proteins of cytoplasmic 14S mRNP particles. D: Fraction of cytoplasmic proteins eluted from DEAE cellulose (the same conditions as for nuclear proteins). E: Fraction of cytoplasmic proteins eluted from DEAE cellulose by urea-2 M NaCl.

**Table 5.** Hybridization of RNA obtained from Nuclear RNP Particles

| Experimental conditions | RNA fractions | Method of RNP preparation* | DNA† | Hybridization: Hybridized RNA (cpm) | Hybridization: Non-hybridized (cpm) RNA | Hybridization per cent |
|---|---|---|---|---|---|---|
| Non infected cells | 30S RNP particles | I | FL | 95 | 1788 | 5.3 |
| | 30S RNP particles | I | Adenovirus | 10 | 3522 | 0.3 |
| | Nuclear polysome-like complexes | I | FL | 40 | 749 | 5.3 |
| Infected cells | 30S RNP particles | II | Adenovirus | 73 | 2709 | 2.7 |
| | Nuclear polysome-like complexes (80-120S) | II | Adenovirus | 36 | 1232 | 2.9 |
| | 30S particles | I | Adenovirus | 72 | 4156 | 1.7 |
| | 30S RNP particles | I | FL | 136 | 1853 | 7.3 |
| | Nuclear polysome-like complexes (80-120S) | I | Adenovirus | 21 | 1520 | 1.4 |
| | Polysomes | II | Adenovirus | 123 | 7056 | 1.7 |
| | Polysomes | II | FL | 87 | 5940 | 1.4 |

* I–RNP particles obtained from 1.41 g/cm$^3$-peak of CsCl gradient. II–RNP particles obtained from sucrose gradient.
† Quantity of DNA: FL–80 mg; Adenovirus–5 mg.

in the RNA isolated from the 30S-component, from poly-particles, and from the complexes purified by banding in CsCl at buoyant density = 1.41-1.42 g/cm$^3$. Thus, following its synthesis in the cell nucleus, the virus specific mRNA is immediately combined with informofers.

The next question is: what is the fate of informofers after the mRNA is released into the cytoplasm? In the cytoplasm of infected cells particles with buoyant density 1.41-1.41 g/cm$^3$ have not been found. The label was concentrated in polysomes and in particles with intermediate density (1.45-1.52 g/cm$^3$). Thus the free complexes of dRNA with informofers were absent from the cytoplasm. This, however, does not provide any information about the fate of the informofer itself. For this purpose it is necessary to compare the protein composition of informofers and cytoplasmic particles containing dRNA. This is a difficult task since the cytoplasmic dRNP particles are not available in a pure state but the first step has been taken in experiments (Lukanidin *et al.*, 1972). The proteins of informofers were compared with proteins of mRNP particles isolated from rabbit reticulocyte polysomes after their dissociation with EDTA. It is known that this mRNP particle is represented by homogeneous 14S particles. In contrast to informofers, polysomal mRNP particles contain several protein components. One of the two main bands in polyacrylamide gel electrophoresis has a mobility similar but not identical to that of informatin (Fig. 11). Thus the protein combined with mRNA in polysomes is not the informatin. It seems most probable, that informofers do not leave the cell nucleus and participate only in the nuclear step of dRNA transport, in accordance with our model (see Fig. 2).

It is possible to consider nuclear particles containing dRNA as "inverted viruses": indeed they consist of homogeneous protein sub-units with RNA distributed at the surface of the structure. This reflects the different function of these particles: in the virus, the capsid protects the RNA; in the nuclear particles the RNA should be attacked by processing enzymes and the protein globule may provide the supporting surface for this. Finally, in both cases an injection of RNA occurs, either from the particles into the cell (viral RNA) or into the cytoplasm (mRNA) without transfer of the protein moiety.

In conclusion, it should be pointed out that although many points in the models presented in Figs 1 and 2 are still speculative, the experiments developed to check them have proved some of the important predictions which follow from the models.

## REFERENCES

Ananieva, L. N., Kozlov, Yu. V., Ryskov, A. P. and Georgiev, G. P. (1968). *Molec. Biol. (U.S.S.R.)* 2, 736.

Arion, V. Ya., Mantieva, V. L. and Georgiev, G. P. (1967). *Molec. Biol. (U.S.S.R.),* **1**, 689.

Bale, W. F., Helmkamp, R. W., Davis, T. P., Izzo, M. J., Goodland, R. L., Contreras, M. A. and Spar, I. L. (1966). *Proc. Soc. exp. Biol. Med.* **122**, 407.

Britten, R. J. and Kohne, D. E. (1968). *Science, N.Y.* **161**, 529.

Church, R. B. and McCarthy, B. J. (1968). *Biochem. Genet.* **2**, 55.

Dreus, J. (1969). *Eur. J. Biochem.* **9**, 263.

Georgiev, G. P. (1969). *J. theor. Biol.* **25**, 473.

Georgiev, G. P. and Mantieva, V. L. (1962). *Biokhimiya,* **27**, 949.

Georgiev, G. P. and Samarina, O. P. (1971). *Adv. Cell Biol.* **2**, 47.

Hatlen, L. E., Amaldi, F. and Attardi, G. (1969). *Biochemistry,* **8**, 4989.

Leppla, S. H., Bjoraker, B. and Bock, R. M. (1968). *In* "Methods in Enzymology", Academic Press, New York and London, XII-B, 236.

Lukanidin, E. M., Georgiev, G. P. and Williamson, R. (1972). *FEBS Letters* (submitted).

Maitra, U. and Hurwitz, J. (1965). *Proc. natn. Acad. Sci. U.S.A.* **54**, 815.

Mantieva, V. L. and Arion, V. Ya. (1969). *Molec. Biol. (U.S.S.R)* **3**, 294.

Mantieva, V. L., Avakyan, E. R. and Georgiev, G. P. (1969). *Molec. Biol., (U.S.S.R.)* **3**, 545.

Olsnes, S. (1970). *Biochim. Biophys. Acta* **213**, 149.

Parsons, J. T., Gardner, J. and Green, M. (1971). *Proc. natn. Acad. Sci. U.S.A.* **68**, 557.

Roblin, R. (1968). *J. molec. Biol.* **31**, 51.

Ryskov, A. P. and Georgiev, G. P. (1970). *FEBS Letters,* **8**, 186.

Samarina, O. P., Asriyan, I. S. and Georgiev, G. P. (1965). *Nauk SSSR* **163**, 1510.

Samarina, O. P., Lukanidin, E. M., Molnar, J. and Georgiev, G. P. (1968). *J. molec. Biol.* **33**, 251.

Takanami, M. (1966). *Cold Spring Harb. Symp. quant. Biol.* **31**, 611.

Williamson, R. and Morrison, M. Paul (1970). *Biochem. biophys. Res. Communs.* **40**, 740.

Zalmanzon, E. S., Lukanidin, E. M. and Mikhalilova, L. N. (1972). *Molec. Biol. (U.S.S.R.),* in press.

FEBS Symposium, Volume 24, 1972, pp. 77-89

# RNA Synthesis in Giant Chromosomal Puffs and the Mode of Puffing*

CLAUS PELLING

*Max Planck Institut für Biologie, Tübingen*
*West Germany*

## MORPHOLOGY OF PUFFING

The polytene chromosome is the only exception to the rule that in differentiated cells chromosomes cannot be visualized when they are in their physiologically active stage, the so-called interphase. This chromosome type is characterized by the formation of specific reversible modifications of chromosomal fine structure (Beermann, 1952), which have been called puffs. Chromosomal RNA synthesis is restricted to these structures. The term puff implies that a particular chromosomal region significantly increases in volume and concomitantly loses much of its structural definition, in contrast to the unpuffed segments which clearly maintain the typical appearance of a specific pattern of alternating bands and interbands. In special cases the alteration of chromosome structure during puffing can reach a drastic extent. The formation of the so-called Balbiani rings, in particular, do not only cause local swelling or partial disintegration of a small region, but involve a complete displacement of the RNA-synthesizing DNA templates relative to their original location. The DNA can be described as being lifted from the chromosomal axis towards the surface of two hemispherical structures which protrude from the chromosomal axis over a considerable distance (often tens of microns) (Fig. 1).

During the process of puffing, the area in which, by autoradiographic means, synthetically active DNA can be shown to be present increases by orders of magnitude; so, the extensive puffing cannot be explained other than by postulating the uncoiling of the deoxyribonucleoprotein fiber. The DNA by virtue of which a puff can extend that far is certainly derived from the DNA located in the chromosomal bands. Interband regions do not contain enough material to feed an unraveling mechanism (see Swift, 1962). The process of uncoiling puts the formation of Balbiani rings and puffs into a

*This contribution is another version of an article prepared for the series: "Results and Problems in Cell Differentiation," Vol. 4, Springer-Verlag, Heidelberg.

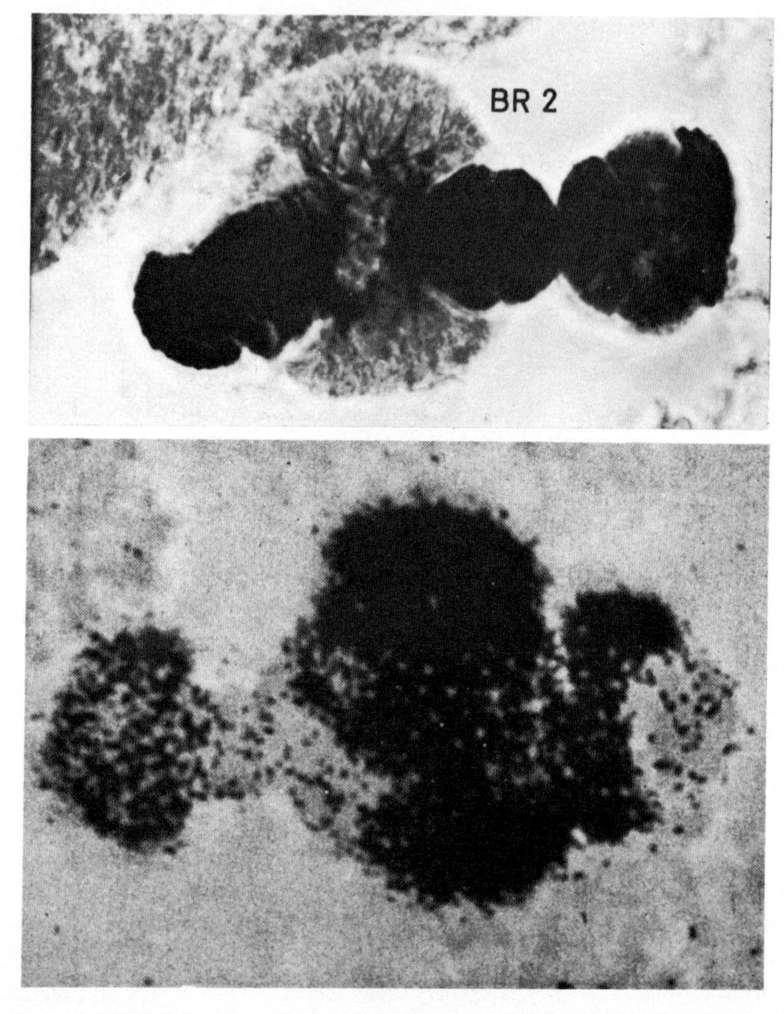


**Figure 1.** Balbiani rings on chromosome IV in a salivary gland cell of *Chironomus tentans*. (a) Carmine-orcein squash preparation, Br 2 indicated. Magnification ca. 900 x (b) RNA specific label in the autoradiograph, magnification ca. 800 x. (From Pelling, 1964.)

context with the famous loop formation in lampbrush chromosomes (Callan and Lloyd, 1960).

These chromosomes, which in cytological terms belong to a completely different chromosome category (they are specialized meiotic chromosomes of the diplotene stage), exhibit the unwinding of chromosomal DNA in connection with the synthesis of RNA more clearly than in giant chromosomal puffs, where the multitude of chromosomal elements building up the polytene chromosome prevents the lightmicroscopical analysis of individual RNA-synthesizing threads. Although cytological analysis indicates that very large puffs physically include series of bands and interbands, it is nevertheless highly probable that the DNA segment properly involved in RNA synthesis is not larger than the DNA of an individual band; it could be demonstrated that giant puffs, such as the aforementioned Balbiani ring or the nucleolus organizer, regularly originate by the activation of one band only. Thus, in their initial stages the giant puffs exhibit the same local specificity which the majority of the puffs never lose due to their small size. It seems therefore consequent to subdivide the DNA of very large puffs into two classes according to their functional significance. A fraction of the DNA seems to be synthetically active. This kind of DNA would be more or less equivalent to the DNA of the band which initiated the puff. The rest of the DNA of the relevant puff is interpreted as inactive and stems from adjacent chromosomal regions which were passively disintegrated in the course of the extensive growth of the puff. Balbiani ring preparations frequently exhibit series of fragmentary and obviously inactive bands, the patterns of which still indicate from which chromosomal regions they have been derived. Quantitatively, the active portion of the DNA in a single chromatid of the bundle constituting the puff would then correspond to the DNA amount of the average chromomere. In *Chironomus*, for example, the number of chromomeres, i.e. bands, approaches 2000 and the DNA of the haploid genome amounts to $2 \times 10^{-13}$ g, according to Edström (1964). The DNA amount of an average chromomere is $10^{-16}$ g, which would be the equivalent of a molecular weight of 60 million daltons.

It may be worthwhile to point out that changes in the extent of puffing, so far discussed in terms of a mechanism which gradually uncoils the deoxyribonucleoprotein fiber, can also be correlated to changes in the RNA-synthetic properties of a puff. Very directly it can be observed that the RNA amount at the puff increases with its increase in size. By autoradiographic analysis this increase seemed to be due to an increase in synthetic capacity of the puff; if the same genetic site is compared in very much different degrees of puffing the uptake of tritiated uridine may easily differ by more than two orders of magnitude. It is this ability to synthesize tremendously varying amounts of RNA which is one of the most typical features of a puff. We will come back to this regulatory flexibility of puffs in an attempt to explain it in molecular terms.

## ACTIVE VERSUS INACTIVE BANDS

Analyses of puff patterns provide information about the ratio of active to inactive chromosomal bands in a differentiated cell. The majority of bands (85%) remain inactive. Such a statement, of course, reaches no further than the limit of resolution of the cytochemical and autoradiographical methods for detecting active structures. In *Chironomus tentans* salivary gland cells there are 1700 unpuffed bands in a total of about 2000 documented bands (Beermann, 1952). The 300 active bands can be further categorized: some of them are characterized by their appearing only at specific times of larval development; others may be present in a particular cell type only. Even the proportion of puffs which do not exhibit any kind of obvious specificity—those which are present most of the time and in each tissue investigated—are not uninteresting; they may be interpreted as maintaining those general metabolic functions upon which every cell type continuously depends. Some quantitative data about the tissue specificity have been obtained by comparing puff frequencies in different tissues of *Drosophila hydei* (Berendes, 1966). In this organism only 5% of all puffs seem to be strictly tissue specific in the sense that they occur in one cell type only. In *Chironomus* 5% of all puffs would constitute a fraction of 0.75% of all chromosomal bands. The size of this fraction in turn gives an idea of how large the genome of *Chironomus* is in terms of the number of different tissues: the pool of the 85% of silent chromosomal bands would be used up when 110 additional cell types had been furnished with their appropriate tissue specific genetic information.

## THE LAMPBRUSH CHROMOSOME LOOP AS A MODEL FOR TRANSCRIPTION

The electron microscope work of O. L. Miller and his co-workers provided an excellent model for the transcription process in higher organisms. Dr. Miller was able to specifically isolate and visualize the transcriptory apparatus of specific chromosomal regions in salamander oocytes. By his methods the spatial relationships between DNA, RNA-polymerase and the transcription product could be studied *in situ.* Figure 2 illustrates ribonucleoprotein fibrils which are attached to a central DNA axis of a lampbrush chromosome loop. Their length increases unidirectionally, thus repeating in the electron microscope the same asymmetry ("polarization") on a molecular level which is also an essential feature of the lampbrush loop in the light microscope (Callan and Lloyd, 1960). The RNA molecules produced are of tremendous length. Although they appear shorter in comparison to the length of the DNA section from which they must have originated, thus having already undergone a process of contraction (Miller *et al.*, 1970), lengths of tens of microns have been reported. This situation makes it highly likely that the DNA in lampbrush loops operates for transcription in very long segments, perhaps as a single uninterrupted strand (Fig. 3).

Remarkable also is the tremendous amount of RNA visible on the loop. It becomes even more striking when described in quantitative terms. The amount

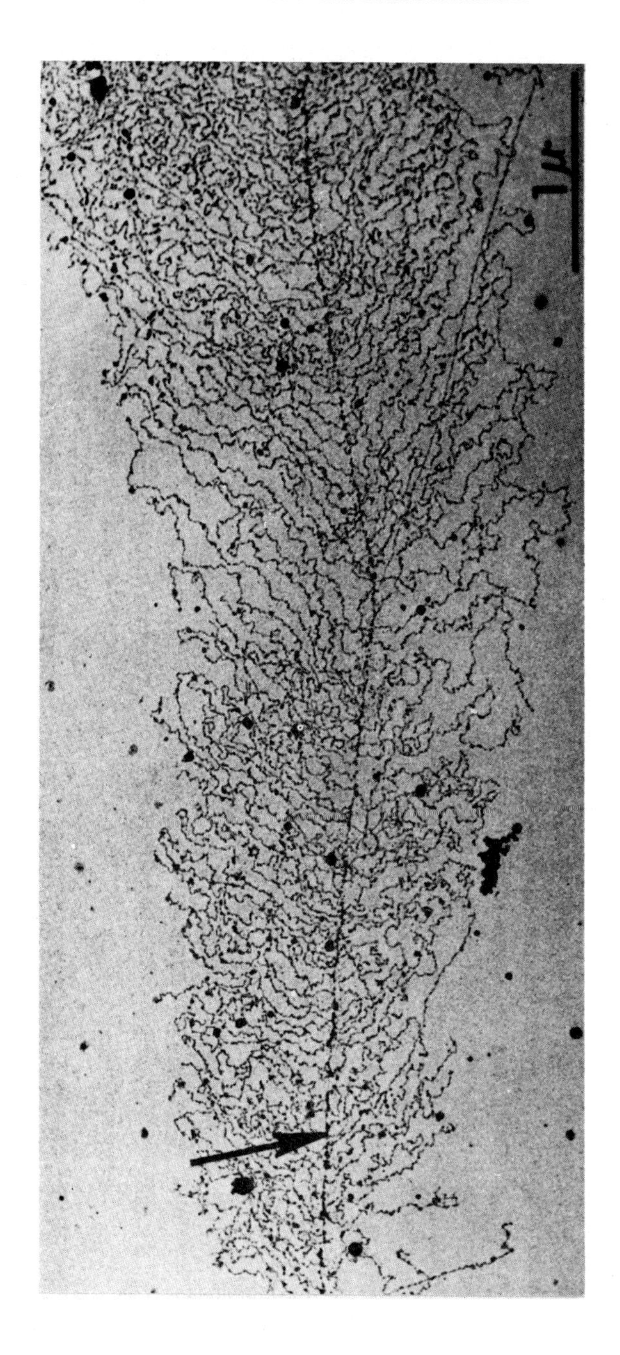


**Figure 2.** Electron micrograph of a portion of a lampbrush chromosome loop isolated from a *Triturus viridescens* oocyte showing the DNA loop axis (arrow) with its gradient of attached RNP fibrils. (Courtesy of Dr. O. L. Miller, Jr.)

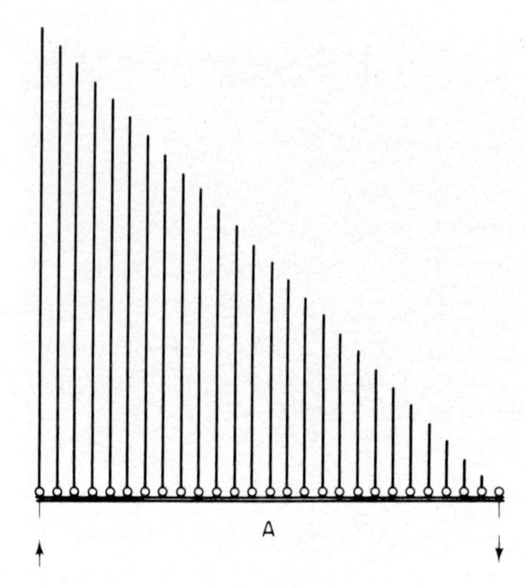


**Figure 3.** Model of transcription derived from a lampbrush loop, entire DNA constituting one unit of transcription.
— DNA, | RNA, ○ Polymerase, ↓ Initiation site ↑ Termination site

of RNA is directly correlated with the density of polymerases along the DNA axis, which in the illustration presented is about 40 polymerases per micron of DNA. On an average loop of 30 micron length this would multiply to 1200 polymerases per loop, each of which would be making an RNA molecule. Provided that the loop contains no more than one unit of transcription, the mean length of an RNA molecule equals half of the loop length. Thus, the total amount of RNA attached to the loop would be 1200 molecules of an average length of 15 microns. In other words, the RNA of a loop exceeds the amount of DNA by a factor of 300.

As a general quantitative expression which also takes into account the possibility that there can be more than one DNA segment active in transcription within the loop, we can write:

$$g_{RNA} = \frac{l_{DNA}^2}{2n_T} \times \rho P \times \rho_{RNA}$$

The RNA amount depends directly on the half of the square of the DNA length ($l_{DNA}$) times the number of polymerases per unit element ($\rho^P$) times the weight of the RNA per unit element ($\rho_{RNA}$). It depends reciprocally on the number of transcriptory elements ($n_T$). Regarding the number of transcriptory elements in a loop, this relationship is maintained only if the size distribution of the elements is not very uneven.

It is possible to derive an estimation of the RNA amount of an average puff from the earlier microelectrophoretic work of Edström and Beermann (1962): apparently a long chromosome such as chromosome I in a salivary gland cell of *Chironomus tentans* contains as much RNA as one single large Balbiani ring on chromosome IV (20 pg). The RNA in chromosome I, however, is distributed over approx. 70 puffs. The RNA amount of an average puff is therefore on the order of 0.3 pg. Taking into account that the giant chromosome of *Chironomus* is a multiple structure containing thousands of identical chromatids (4000 or 8000 in most nuclei) quite different RNA to DNA ratios are obtained in comparison to the lampbrush loop: in a Balbiani ring there exists only 33 times more RNA than DNA. In a puff of average size, however, the DNA even exceeds the RNA by a factor of 2. Provided that the whole chromomeric DNA of a puff is involved in transcription and that $n_T$ equals 1 then it follows that an ordinary puff operates with only two polymerases per strand. Even the gigantic Balbiani ring which is most certainly the largest puff structure seen in any Dipteran cell contains about 120 polymerases per strand.

## DIFFERENCE BETWEEN THE LAMPBRUSH AND THE POLYTENE CHROMOSOME SYSTEMS

Polymerase density apparently differs between the lampbrush chromosome loop and the giant chromosome puff by at least one (Balbiani ring), and frequently by more than two, order of magnitude (puffs). It is also relevant to point out that one of the two systems, the lampbrush loop, seems to be so densely packed with polymerases that there is little free space available to put further molecules on the DNA strand. As Miller makes clear in a consideration of polymerase distribution on nucleolar genes in amphibia, a physical separation of polymerase molecules by a distance of 100 to 200 Å approaches saturation (Miller and Beatty, 1969), since the diameter of the molecules is in the order of 100 Å (Lubin, 1969). This feature of the lampbrush transcriptory system, on the one hand, seems to be an advantage for cells which require a high output of transcriptory products, but on the other, throws doubt on the efficiency of the regulatory capacity of such RNA synthesis. In this respect, saturation of the DNA with polymerases is in concert with two other characteristics of the lampbrush chromosome: First, in lampbrush chromosomes, seemingly all chromomeres (not all DNA!) are activated along the entire length of the chromosome. Second, the loop formation is a gradual, long term process which seems to involve all the active structures in an almost identical way. This indicates a fairly low degree of differentiation between the loops.

The giant chromosomal system is very much different. Only a small and selected portion of genetic sites is activated, and these structures exhibit rapid and individual responses (puff induction and regression) to many different kinds of external and internal stimuli.

It seems that the transcriptory system of the oocyte is inflexible in terms of gene regulation at the chromosomal level, at least in comparison to the giant chromosome. An understanding of the biological significance of this difference currently seems impossible. Qualitatively, the transcription of such a large amount of presumably diverse genetic information should not, in an oocyte, have a meaningful realization at the phenotypic level. The uniformity of transcription in a system, the most obvious purpose of which has to be growth and the making available of substrates for extensive cell division, may suggest a widely unspecific use of the RNA synthesized. One may conceive also of an unconventional explanation: the ribosomal RNA synthesis might depend on a concomitant production of RNA of the nuclear heterogenous type. The intensity of the nucleolar RNA synthesis in the amphibian oocyte nucleus which has reached the state of extreme gene amplification would then be reflected in the turning on of very many other genes.

## PUFFING AS A CONSEQUENCE OF VARYING CONCENTRATIONS OF RNA-POLYMERASE

As already mentioned, the difference in the mount of RNA does not only distinguish different categories of active loci, it also characterizes different degrees of puffing at a single genetic site. The same assumptions upon which we have based our calculations of polymerase concentrations of different loci can sensibly be applied also in this case: unless there is a change in the number of transcriptory elements, or in the length of the DNA segment transcribed which changes the products of the puff during the process of uncoiling, only the polymerase density determines the RNA amount at the puff level. Although the influence of puffing on the nature of the products of one puff has not been reliably analysed, it seems clear that the tremendous difference in RNA content of a puff can be most easily explained as being due to a change in polymerase concentration in the DNA fiber. This seems all the more justified since the RNA synthetic capacities of a puff parallel the differences in its content of RNA. Furthermore, the morphological change of the DNA within the band during puffing can be understood on the basis that the progress of a single RNA polymerase molecule along the chromomeric DNA should not cause any microscopically obvious change in the structure of the chromosomal band. Complete lack of visual alteration in giant chromosomal structure during the action of DNA polymerases may serve here as a comparison (Pelling, 1969).

Only the loading of the DNA fiber with a multitude of RNA polymerases seems to have sufficient morphological consequence so that a puffing phenomenon could be recognized in the light microscope. An increase of space requirement would have to follow the accumulation of polymerases due to the increase of RNA present. Moreover, uncoiling of the DNA could also be caused by the increasing rigidity which a decreasing distance between the polymerases would impose on the DNA segment, thus preventing the maintenance of any tight coiling. Whether the role of the polymerase molecules is overemphasized by such a model, which claims the polymerases to be the determinant of the extent of puffing, may be tested; the two essential parameters, initiation and termination, by which the polymerase concentration on the template may be regulated, can be experimentally varied. If, for example, the initiation process is blocked, possibly by rifampicin analogues (Butterworth *et al.*, 1971) without otherwise affecting the transcriptory mechanism, the template would run out of polymerases and shrinkage of puffs would be the consequence. Selective inhibition of the termination process, on the other hand, would largely induce the reverse phenomenon. The puff DNA could become saturated with polymerases, thus enlarging the active chromosomal structure in question. An inhibition of elongation would lead to a standstill of the system, and would also preserve the status quo morphologically, provided that the polymerases remain attached to the template.

A final aspect may be emphasized. Experiments of Berendes (1968) describe a considerable accumulation of acidic proteins immediately after puff induction in the presence of actinomycin. This suggests that a preinduction process would have to be distinguished from the uncoiling process proper. The latter might be caused by the elongating polymerases; the former, which may or may not include the binding of the polymerase molecules to the template together with the initiation process, may have something in common with the locus specific processes in puff formation.

## LENGTH OF THE RNA MOLECULES TRANSCRIBED IN A PUFF

Our considerations have so far been based on two main assumptions, the first being that the transcribing system of lampbrush loops and giant chromosome puffs consists of a single continuously transcribing segment, and the second that the whole DNA of the particular structure participates in the transcription. No evidence has so far been presented that these assumptions are correct in the case of the giant chromosome system, and it is clear that the above conclusions on the significance of polymerase concentrations for the process of puffing remain imaginary whilst the fundamental similarity between loops and puffs cannot be substantiated. In such a case the transcriptory

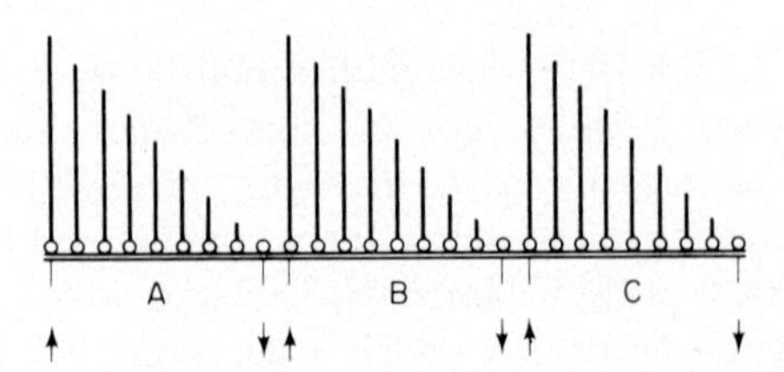


**Figure 4.** Mode of transcription, similar to Fig. 3. DNA is subdivided into 3 transcriptory elements of equal length. For symbols see legend of Fig. 3.

system of a puff could reach any degree of complexity; the length of DNA involved in RNA synthesis could be smaller, and the number of transcriptory elements could be numerous. According to the relationship on page 82 the number of polymerases on the template, then, would have to increase in order to account for the loss of RNA on the relevant structure. Figure 4 illustrates a DNA segment subdivided into three identical transcriptory elements. Compared to the situation demonstrated in Fig. 3 the subdivision of the DNA leads to a decrease of RNA by a factor of 3. The difference in RNA amount could also be accounted for in different ways. We could envisage the existence of a pool of finished molecules within a puff. These molecules may not have had the opportunity to leave the puff area immediately after their completion.

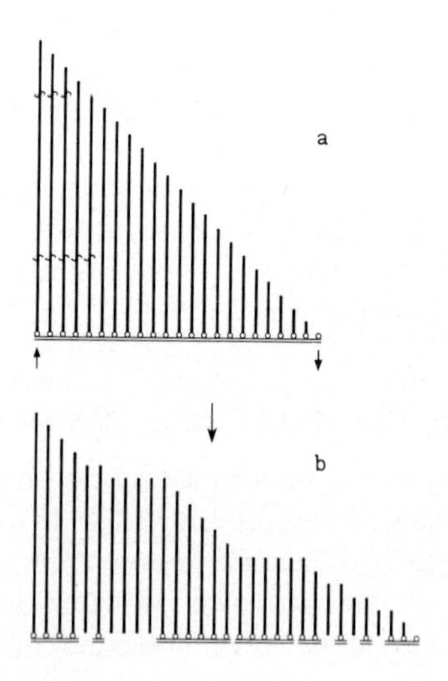


**Figure 5.** Mode of transcription in a puff. RNA is processed at the chromosomal level (a) situation of the puff transcriptory unit; (b) array of molecules according to size, indicating molecular heterogeneity of the system. ~ cleavage point, for other symbols see legend of Fig. 3. (Changed after Pelling, 1970.)

An even more complex puff model has been suggested in order to explain the simultaneous presence of large and small RNA molecules at a puff in earlier analyses (see below). This model (Fig. 5) postulated that a Balbiani ring is constructed of a single transcriptory unit as in the model discussed above (Fig. 4a). But contrary to this simple version of the model, the ordered relationship between the length of DNA, the concentration of polymerases and the RNA amount is altered by a mechanism which processes the RNA molecule on the RNA template, sometimes before its synthesis has come to an end. This modified model explains the minuteness of the RNA amount at a puff, not by assuming a very low polymerase concentration as we have done before, but by invoking the action of a process which chooses the earliest possible moment to transport the RNA away from the template. This model would, at times, place the mechanism of molecular processing, which has to be postulated in order to account for the difference in molecular weight between nuclear heterogenous RNA and cytoplasmic "messenger RNA", in the immediate neighborhood of the chromosomes.

Fortunately there is information concerning the length of puff RNA molecules and this in turn provides information about the transcriptory properties of the active site. We know from earlier analyses of nuclear heterogenous RNA (Scherrer *et al.*, 1966) that the average molecular weights are quite high, in the order of several millions. In terms of the number of transcriptions, we could deduce from these data that we would not have to

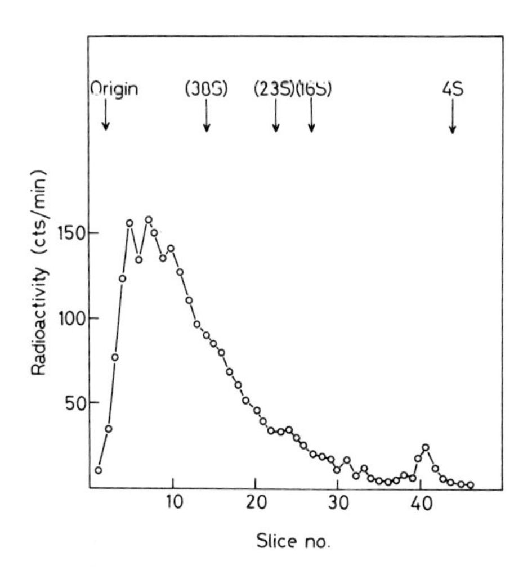


**Figure 6.** Electrophoretic analysis of labeled RNA from BR II of *Chironomus tentans.* Salivary glands were incubated for 45 min in larval hemolymph, provided with tritiated cytidine and uridine. Fifty BR II were isolated, RNA extracted and analysed by electrophoresis in 2% agarose. (Courtesy of Dr. B. Daneholt, Stockholm.)

expect more than a few transcribing elements per chromomere provided that the duck erythroblast system were largely comparable to the *Chironomus* system discussed here.

In the past years data on the size of RNA molecules from giant chromosomes and single active loci have been obtained (Daneholt *et al.*, 1970; Pelling, 1970). The main feature of these earlier RNA separations, one of which is illustrated in Fig. 6, is the high degree of complexity of the RNA. The most straightforward interpretation of such a picture regards the multitude of the RNA types as a reflection of the complexity of the transcriptory elements within a single puff region. Apart from the heterogeneity of the patterns, however, there is another characteristic of the profiles to be accounted for. These patterns obtained from Balbiani rings were not constant in themselves. Sometimes the radioactivity focused rather on the left side of the spectrum where the particularly large molecules are expected to be localized, sometimes it shifted more to the right. This quasi-kinetic behaviour of the RNA spectra prevented any simple explanation as to the transcriptory nature of a puff (Daneholt *et al.*, 1970; Pelling, 1970).

This was the situation until very recently when Daneholt demonstrated that when the RNA isolation procedure is carried out with special precaution against the action of enzymic degradation, RNA of Balbiani ring 2 of *Chironomus tentans* can be isolated as one uniform peak (Daneholt, in press). The aforementioned heterogeneity, therefore is rendered likely to have originated from a breakdown–though slow, and judging from the reproducibility of peaks in the profiles published, not unspecific breakdown–of RNA during micromanipulation of chromosomes and during the RNA isolation procedure. This peak, representing an RNA which has taken up its label during 90 minutes of an *in vitro* incubation of the salivary gland cells with tritiated RNA precursors, has a sedimentation coefficient of more than 100S. Its molecular weight is, according to Daneholt, in the order of tens of million daltons. Such a large molecule clearly reflects the tremendous length of the transcribing segment of the DNA within the Balbiani ring. It corresponds to the 60 million daltons which have been calculated for the total DNA amount of the average chromomere. With this finding also, the comparability of the lampbrush loop, from which we started our consideration, and the Balbiani ring has been sufficiently established. Far from claiming to have understood the chromosomal transcriptory apparatus in its details we have, however, proceeded to substantiate the concept of the fundamental compartmentalization of the chromosome of higher organisms presented earlier (Beermann, 1966; Pelling, 1966). The chromomeres, the linear array of which builds up the chromosome structure proper, have been postulated to be units of replication and transcription. It appears now, with respect to transcription, that this postulate is maintained even in a more direct sense than one could have expected in molecular terms.

## ACKNOWLEDGEMENTS

I am particularly indebted to Dr. B. Daneholt for providing me with some still unpublished results and with Fig. 6; I am very grateful to Dr. O. L. Miller, Jr. for his permission to use again the electron micrograph of the lampbrush loop and to Dr. P. Rae for invaluable help in preparing this manuscript. I also wish to thank Drs. K. Götz and H. Stein for helpful discussion, Frl. H. Bürgermeister for most careful technical assistance and Herrn E. Freiberg for executing the illustrations.

## REFERENCES

Beermann, W. (1952). *Chromosoma* **5**, 139–198.
Beermann, W. (1961). *Chromosoma* **12**, 1–25.
Beermann, W. (1966). *In*: Cell Differentiation and Morphogenesis pp. 24–54, North Holland Publishing Company, Amsterdam.
Berendes, H. D. (1966). *J. exp. Zool.* **162**, 209–217.
Berendes, H. D. (1968). *Chromosoma* **24**, 418–437.
Butterworth, P. H. W., Cox, R. P. and Chesterton, C. J. (1971). *Eur. J. Biochem.* **23**, 229–241.
Callan, H. G. and Lloyd, L. (1960). *Phil. Trans. Roy. Soc.* **B.243**, 135–219.
Daneholt, B., Edström, J.-E., Egyhazi, E., Lambert, B. and Ringborg, U. (1970). *Cold Spring Harb. Symp. quant. Biol.* **23**, 513–519.
Edström, J.-E. and Beermann, W. (1962). *J. biophys. biochem. Cytol.* **14**, 371–380.
Edström, J.-E. (1964). *In*: Role of Chromosomes in development, p. 137, Academic Press, New York and London.
Gall, J. G. (1963) *In*: Cytodifferential and Macromolecular Synthesis, p. 119–143, Academic Press, New York and London.
Grossbach, U. (1969). *Chromosoma* **28**, 136–187.
Keyl, H.-G. (1969). *J. molec. Biol.* **39**, 219–233.
Lubin, M. (1969). *J. molec. Biol.* **39**, 219-233.
Miller, Jr., O. L. and Beatty, B. R. (1969). *Science, N.Y.* **164**, 955–957.
Miller, Jr., O. L., Beatty, B. R., Hamkalo, B. A. and Thomas, Jr., C. A. (1970). *Cold Spring Harb. Symp. quant. Biol.* **25**, 505–512.
Pelling, C. (1964). *Chromosoma* **15**, 71–122.
Pelling, C. (1966). *Proc. Roy. Soc. B,* **164**, 279–289.
Pelling, C. (1969). *Progress Biophys. Mol. Biol.* **19,1**, 239–270 Pergamon Press, Oxford.
Pelling, C. (1970). *Cold Spring Harb. Symp. quant. Biol.* **25**, 521–531.
Scherrer, K., Marcaud, L., Zajdela, F., London, J. M. and Gros, F. (1966). *Proc. natn. Acad. Sci. U.S.A.* **56**, 1571–1578.
Swift, H. (1962). *In* The Molecular Control of Cellular Activity, pp. 73–125, McGraw-Hill, New York.

FEBS Symposium, Volume 24, 1972, pp. 91-98

# The Ribosomal Genes During Amphibian Oogenesis

MARCO CRIPPA*

*C.N.R. Laboratory of Molecular Embriology*
*Arco Felice, Italy*

and

GLAUCO P. TOCCHINI-VALENTINI

*C.N.R. Laboratory of Cell Biology*
*Rome, Italy*

The accumulation by the amphibian oocyte, at the end of a long and very active synthetic period, of a large amount of ribosomes can be taken, with certain limitations, as an example of a differentiated cellular function. The oocyte in this respect behaves like a cell which at a certain time of its life cycle has adapted itself in order to become very efficient in the synthesis and in the accumulation of a specific gene product, in this case ribosomal RNA. Since ribosomal RNA can be easily recognized and studied and since the sequences which are specifying it can be obtained in pure form, the analysis of the ribosomal system has proved useful in understanding some molecular mechanisms which are involved in the regulation of the expression of the ribosomal genes in higher organisms.

If we consider how the oocyte responds to the need of supplying ribosomal RNA to a mass of cytoplasm which normally would contain several thousand nuclei, we can recognize different moments of this response. Initially, the oocyte increases the amount of template available for ribosomal RNA synthesis by increasing over a thousand fold the number of its ribosomal genes. As a result of this selective amplification process, more than 50% of the DNA contained in the oocyte nucleus consists of extrachromosomal copies of the ribosomal genes.

Later during oocyte growth these genes are very actively transcribed. From the beautiful pictures of Miller and Beatty (1969) it appears that all the

* Present address: Department of Animal Biology, University of Geneva, Geneva, Switzerland.

ribosomal cistrons are engaged in transcription at the maximum possible rate and that the growing RNA chains are very tightly packed. The transcription of the ribosomal genes reaches its maximum at a time when the chromosomes are in the lampbrush stage and subsequently decreases until, just prior to maturation, all of the ribosomal RNA synthesis has almost stopped, even though the extrachromosomal copies of the rRNA cistrons are still present in the oocyte nucleus. We will separately consider these different moments.

## AMPLIFICATION

Recent evidence from analysis of rDNA amplification in interspecies hybrids suggests that the amplification of the ribosomal genes occurs through a chromosome copy mechanism (Brown and Blackler, 1972), i.e the first amplified copy is made on one of the several hundred repeating gene sequences which are present in the nucleolar organizer region of the chromosome. The existence of a chromosome copy mechanism poses important questions since the cell is expected to have enough specificity in its DNA replication mechanisms to be able to recognize and amplify a very small part only (0.2%) of its genome. We recently obtained data showing that this specificity is ensured by transcribing first a complete rDNA unit into an RNA molecule and then utilizing this RNA transcript for RNA dependent DNA synthesis (Crippa and Tocchini-Valentini, 1971). When ovaries taken from young animals are incubated in a medium containing radioactive thymidine, the DNA is extracted and centrifuged to equilibrium in a cesium chloride

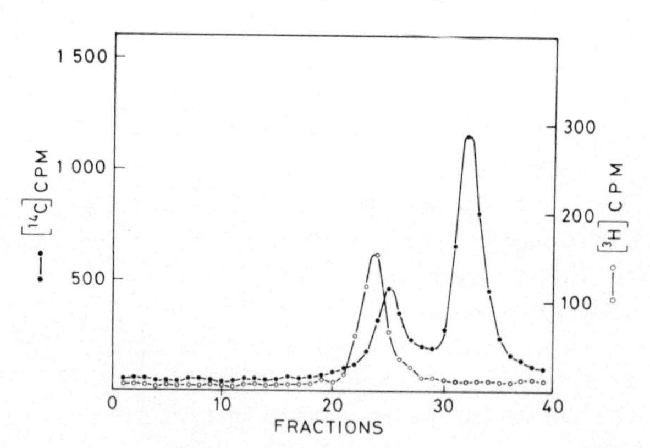


**Figure 1.** Equilibrium density centrifugation of labeled nucleic acids extracted from ovaries during the amplification stage. The ovaries were incubated for 48 h with 50 $\mu$Ci/ml of $^{14}$C-thymidine (Amersham, 400 Ci/mol) and with 400 $\mu$Ci/ml of $^{3}$H-uridine (Amersham, TRK). The extracted nucleic acids were incubated with 20 $\mu$g/ml of boiled pancreatic RNase for 1 h at 37°C, subsequently reextracted with phenol, alcohol precipitated, and resuspended in 10 mM Tris-HCl (pH 8.1) for the CsCl centrifugation.

density gradient, it is possible to monitor the synthesis of the amplified ribosomal cistrons. These band at a density of 1.729 gm/cm$^3$, clearly separated from the bulk chromosomal DNA which bands at a density of 1.700 gm/cm$^3$ (Gall, 1968). If the ovaries are incubated in a medium containing both radioactive thymidine and radioactive uridine, an RNA component banding associated with the heavy region of the amplified ribosomal cistrons peak is detected (Fig. 1).

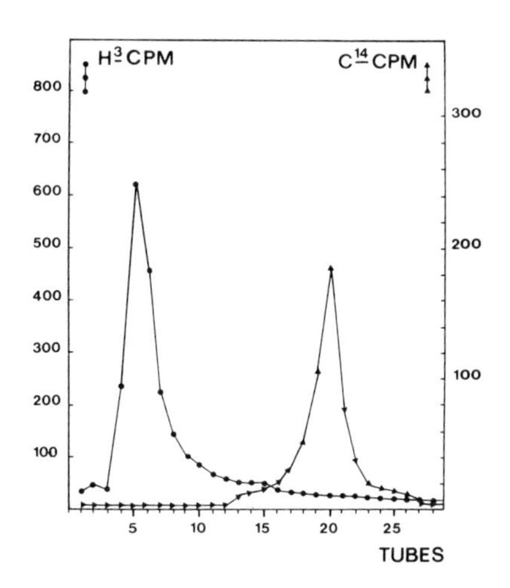


**Figure 2.** Equilibrium density centrifugation of the denaturate RNA rDNA complex in Cesium sulphate. The nucleic acids were labeled and centrifuged as described in the legend of Fig. 1 but the RNA'se treatment was omitted. The heavy region of the rDNA peak was taken, dialysed against 10 mM Tris pH 7.6-10 mM NaCl; heat denaturated, adjusted to 1 mM EDTA and centrifuged to equilibrium in Cesium sulphate (70 h at 33,000 rpm, 23°C, in a SW 50.1 Spinco rotor). ●–●–● $^3$H-uridine counts; ▶–▶–▶ $^{14}$C-thymidine counts.

When the region of the rDNA peak containing the RNA is taken, denatured by heat and banded in a cesium sulphate gradient, the RNA bands at a density of 1.630 gm/cm$^3$, completely separated from the single-stranded ribosomal cistrons (Fig. 2). Similarly, complete separation of the RNA from the DNA can be obtained, after denaturation, by filtration on a nitrocellulose filter. More than 98% of the input RNA counts are recovered as TCA precipitable material in the filtrate whereas virtually all the DNA counts are retained by the filter. From these results it can be concluded that most of this RNA component is associated with the DNA in a complex sensitive to treatments which disrupt hydrogen bonds and it is therefore not convalently linked to the DNA. The nature of the RNA-rDNA complex was further investigated by analysing its size on sucrose gradients before and after ribonuclease treatment (Brown and Tocchini-Valentini, 1972; Mahdavi and

Crippa, 1972). The native complex, prepared from ovaries incubated for 48 h *in vitro* with $^3$H-thymidine, sediments between the 18S and the 28S RNA markers. After ribonuclease digestion the thymidine counts show a dramatic change and can be found to sediment in the 9-10S region of the gradient (Fig. 3).

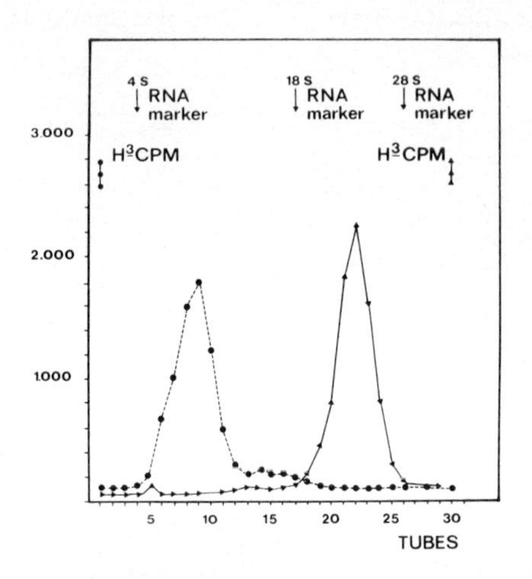


**Figure 3.** Sucrose density of $^3$H-thymidine labeled newly synthesized rDNA. $^3$H-labeled rDNA was extracted from ovaries incubated for 48 h *in vitro* with 100 $\mu$c/ml of $^3$H-thymidine (Amersham, TRK) and purified by Cesium chloride equilibrium centrifugation. The rDNA peak was taken, dialysed against 10 mM Tris pH 7.6-10 mM NaCl-1 mM EDTA, and treated with ribonuclease (preheated Wortington pancreatic ribonuclease A, 100 $\mu$g/ml, 30 min at 37°C). The controls were similarly incubated but without ribonuclease. The samples were then centrifuged for 18 h at 2°C and 25,000 rpm in a SW Spinco rotor, in a 15-30% sucrose gradient. ▶—▶—▶ controls; ●—●—● ribonuclease treated.

When the same kind of experiment is performed on amplified rDNA labeled *in vivo* for 48 h by injecting 10 $\mu$c of $^3$H-thymidine into small metamorphosing frogs, the pattern obtained is completely different. The labeled amplified DNA has a much higher sedimentation coefficient which does not change after ribonuclease treatment (Fig. 4). These results clearly prove that the newly synthesized ribosomal DNA chains are held together in a complex structure by RNA component. *Xenopus* rDNA is made of repetitive units comprising the sequences coding for 28S and 18S RNAs and a portion of higher GC content called spacer. Since the anatomy of the amplified rDNA cistrons is the same as the anatomy of the chromosomal rDNA cistrons, the RNA transcript isolated from ovaries during the amplification stage, in order to be used as template for gene amplification, should also contain the sequences corresponding to the spacer. Figure 5 shows the results obtained in the following experiment. Ovaries from metamorphosing frogs were incubated

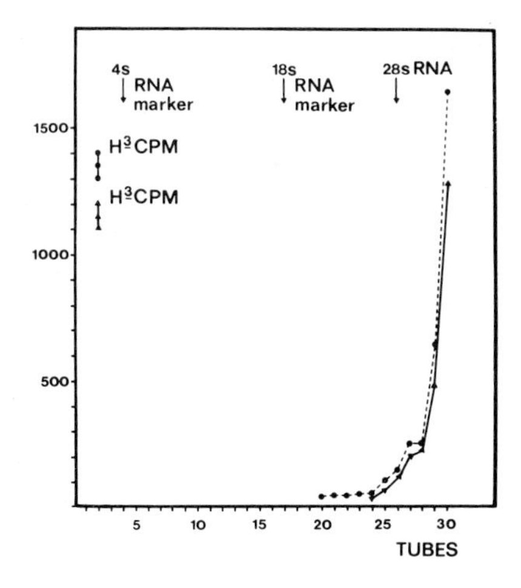


**Figure 4.** Sucrose density analysis of $^3$H-thymidine labeled newly synthesized rDNA. The rDNA was extracted and treated as described in the legend of Fig. 3. The labeling was done for 48 h by injecting 10 $\mu$l of aqueous solution of $^3$H-thymidine (I mc/ml, Amersham) into the abdominal cavity of young metamorphosing frogs. ▶—▶—▶ controls; ●—●—● ribonuclease treated.

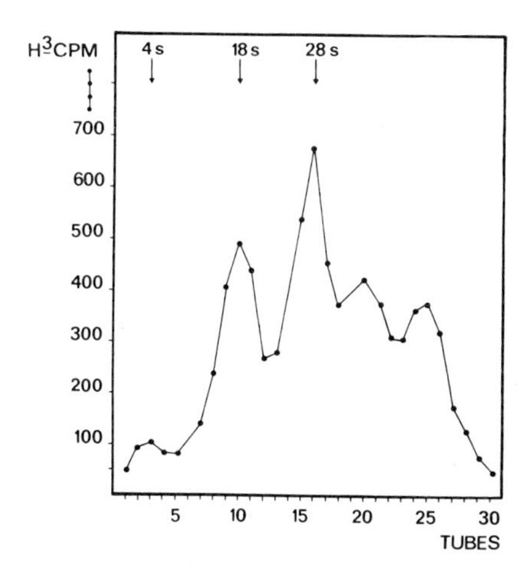


**Figure 5.** Hybridization profile of a sucrose gradient of RNA extracted from ovaries during the amplification stage. The ovaries were incubated for 48 h in a medium containing 400 $\mu$c/ml of $^3$H-uridine (Amersham, TKR), the RNA was extracted and centrifuged in a sucrose density gradient (15-30%) for 20 h at 27,000 rpm in a SW 27 Spinco rotor). Each fraction of the gradient was dialysed against 2 XSSC and annealed with 0.5 $\mu$g of purified rDNA for 20 h at 66°C.

*in vitro* for 48 h in a medium containing 400 μc/ml of 5-$^3$H uridine. The total RNA was extracted and analysed on a sucrose gradient. Each tube of the gradient was then hybridized with purified ribosomal cistrons. In the profile of the hybridized counts one can detect the peaks corresponding to the 18S and 28S ribosomal RNAs, to the 40S ribosomal precursor and to another much faster sedimenting RNA component which has the molecular weight expected for the complete transcript of one repetitive unit of the ribosomal DNA. By hybridization with spacer-rich DNA it was possible to prove that in this RNA molecule, in addition to the 18S and 28S sequences, the spacer sequences were also present.

Recently a similar large RNA molecule (4.4 millions dalton M.W.) has been observed in the liver of three amphibians, including *Xenopus laevis*, by Caston, *et al.* (1972). They interpret this molecule as the real ribosomal precursor.

## TRANSCRIPTION

Shortly after amplification has been completed the oocyte initiates the growth phase which lasts for several months and which allows the accumulation of many gene products, including ribosomal RNA, for later use during development.

Has the oocyte transcription machinery adapted in some way to match these special requirements?

We have previously reported the isolation and the purification from *Xenopus* ovaries of two different RNA polymerase activities; only one of them is localized in the nucleolus. On the basis of localization, binding to purified rDNA and response to inhibitors, we concluded that the nucleolar RNA polymerase is responsible for ribosomal RNA synthesis *in vivo.* This enzyme has now been purified and has the characteristics of the AI enzyme of the Chambon classification.

In *Xenopus* ovaries we have also detected the presence of a fairly large amount of "free" RNA polymerase which is not chromatin bound (or alternatively which is very easily solubilized) and which is therefore lost using the classical Roeder and Rutter (1969) procedure. At the moment it has been impossible to assess the cellular localization of this "free" RNA polymerase activity. This activity elutes from DEAE-cellulose at a slightly higher ionic strength than the nucleolar RNA polymerase and has an intermediate α-amanitin sensitivity. The enzyme becomes sensitive at very high concentration of the drug (30-50 μg/ml) whereas the nucleolar enzyme is completely insensitive even at concentrations of α-amanitin as high as 200 μg/ml.

In order to have a better understanding of the functions and of the specificity of the different cellular RNA polymerases we analysed the *in vitro* transcription of purified ribosomal cistrons by the purified nucleolar enzyme.

Between 75% and 80% of the RNA made *in vitro* is complementary to the strand which is normally transcribed *in vivo*. Since, however, strand selection does not represent such a stringent criterion for judging fidelity of transcription, we chose to investigate the different initiation sites at which transcription was started. The *Xenopus* nucleolar enzyme preferentially initiates with GTP. If the incorporation of γ-labeled $P^{32}$-GTP is taken as

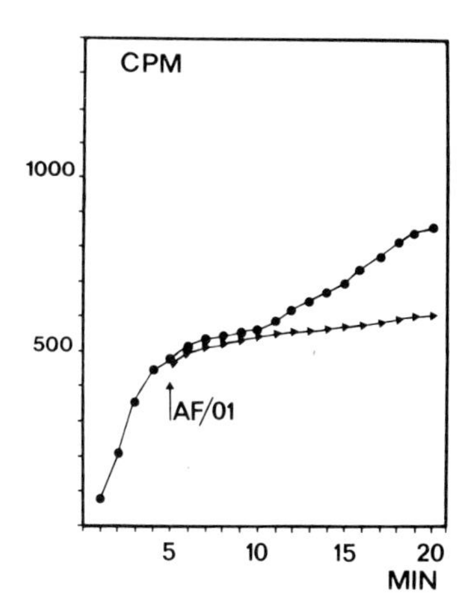


**Figure 6.** Time course incorporation of γ-labeled $P^{32}$-GTP. The reaction was carried out for different times as described previously (Tocchini-Valentini and Crippa, 1970). A 5 to 1 template (purified rDNA)-enzyme (purified *Xenopus* nucleolar RNA polymerase) was used. The γ-labeled GTP was prepared and purified by ourselves and had a final specific activity of 30 millions cpm/nanomole. When indicated, AF/01 was added to a final concentration of 70 μg/ml. ●–●–● $P^{32}$-GTP incorporation; ▶–▶–▶ AF/01 added after 5 min of incubation.

100% reference value, the efficiency of the ATP initiation is about 30% and the efficiency of the pirimidine triphosphates no more than 3-4%. In order to avoid mistakes in initiation due to a lack of release of the newly synthesized RNA chains and consequently to a jamming of the system, the experiments were carried out with a large excess of template (template/enzyme = 5 to 1 molar ratio). Under these conditions the γ-$P^{32}$ incorporation increases linearly for the first 5 minutes of the *in vitro* reaction whereas for longer incubation periods the rate of incorporation tends to decrease (Fig. 6). One can prevent almost completely any further incorporation of labeled $P^{32}$-GTP by adding to the incubation mixture a rifampicin-derivative (AF/01) which has been shown to inhibit eukaryotic polymerases by binding to the free enzyme and blocking in this way initiation (E. Mattoccia, personal communication).

The γ-$P^{32}$-GTP labeled product synthesized *in vitro* after 5 min of incubation was extracted and treated with pancreatic ribonuclease A under conditions of complete digestion. The digestion products were separated and three of them were reproducibly shown to have $P^{32}$ label. Since, however, one of the products is a dinucleotide (pppGpC) it is quite likely that it could derive from the digestion of different sequences all having guanine and cytosine residues at the 5′ terminal end.

Different attempts to decrease the number of initiation sites have so far been unsuccessful. In the same way it was impossible to show specific effects on the pattern of initiation of a factor that we have previously described (Crippa, 1970).

By competition-hybridization experiments it is possible to show that approximately 70% of the initiation takes place in the 18S region of the cistron, 25% in the spacer region and 5.8% in the 28S region.

It seems therefore that the purified chromatin-bound nucleolar enzyme lacks the specificity necessary to recognize on naked rDNA a specific starting signal. The role of different initiation factors can at this point become essential. Indeed in the cell the DNA is normally complexed with different proteins and it is therefore quite possible that one or more of these proteins could be necessary to ensure specificity of transcription.

## ACKNOWLEDGEMENTS

We thank Professors Lancini and Silvestri (Lepetit S.p.A., Milano, Italy) for the generous gift of the rif-derivatives. E. Mattoccia for informing us of his unpublished results on the mechanism of action of these rif-derivatives and S. Lamberti, G. Locorotondo and B. Esposito for expert technical assistance.

## REFERENCES

Brown, D. D. and Blackler, A. W. (1972). *J. molec. Biol.* **63**, 75.
Brown, D. D. and Weber, C. S. (1968). *J. molec. Biol.* **34**, 661.
Brown, R. and Tocchini-Valentini, G. P. (1972). in press.
Caston, J. D., Jones, P. and McIntyre, S. (1972). in press.
Chambon, P. (1971). FEBS Letters, June-July.
Crippa, M. (1970). *Nature, Lond.* **227**, 1138.
Crippa, M. and Tocchini-Valentini, G. P. (1971). *Proc. natn. Acad. Sci. U.S.A.* **68**, 2769.
Gall, J. C. (1968). *Proc. natn. Acad. Sci. U.S.A.* **60**, 553.
Landesman, R. and Gross, P. R. (1969). *Devl. Biol.* **19**, 244.
Loening, U. E., Jones, K. and Birnstiel, M. L. (1969). *J. molec. Biol.*, **45**, 353.
Mahdavi, V. and Crippa, M. (1972). in press.
Miller, O. L. and Beatty, B. R. (1969). *Science, N.Y.* **164**, 955.
Roeder, R. G. and Rutter, W. J. (1969). *Nature, Lond.* **224**, 234.
Tocchini-Valentini, G. P. and Crippa, M. (1970). *Nature, Lond.* **228**, 993.

FEBS Symposium, Volume 14, 1972, pp. 99-107

# The Complexity of RNA Transcription from Single Copy DNA During Mouse Development

R. B. CHURCH and I. R. BROWN

*Divisions of Medical Biochemistry and Biology,*
*Faculty of Medicine, The University of Calgary,*
*Calgary, Alberta, Canada*

The regulation of genetic activity has been elegantly illustrated in a large number of prokaryotic systems (see Cold Spring Harbor Symposium, Vol. 35, 1970). The elucidation of the genetic fine structure in the genome of viruses and bacteria has provided useful models in the explanation of the control of transcriptional patterns of RNA synthesis during the proliferation of these organisms. Regulatory mechanisms have been proposed to explain the process of translation whereby the genetic information carried in the messenger RNA serves as a template for polypeptide synthesis. In most systems examined so far protein synthesis seems to involve an accumulation of stable ribosomal and transfer RNA species and a dynamic utilization of unstable messenger RNA molecules in the cell.

In contrast, the details of cellular mechanisms controlling selective gene expression in animal systems, particularly *in vivo,* remains relatively obscure. The nuclear membrane of eukaryotes compartmentalizes the cell with definitive effects on the transport of unstable nuclear RNA transcripts from the nucleus to the cytoplasm. The kinetics of the rapid turnover of this heterogeneous nuclear RNA in mammalian cell nuclei was first noted by Harris (1963). The differential transport of this unstable nuclear RNA from the nucleus to the cytoplasm in mammalian cells has been studied *in vivo* and *in vitro* by a number of laboratories (see McCarthy *et al*, 1969). The spectrum of messenger RNA transcripts which are selectively transported to the cytoplasm must be responsible for the basic biochemical differences which exist between tissues in an animal. The process of differentiation, therefore, may be described as the process whereby cells, presumably containing identical nuclear DNA sequences, develop into phenotypically distinct entities each with a characteristic pattern of constitutive proteins. The differences in gene activity which initiate distinct cell protein profiles at different stages of development in mammals and in different tissues can involve "regulatory mechanisms" which operate at the level of RNA transcription, selective modification and transport of RNA molecules from the nucleus to the cytoplasm, and differential stabilities of both presumptive messenger RNA and polypeptide products. The pattern of protein molecules characteristic of any given cell type probably involves a dynamic equilibrium representing the summation of regulation at all of the above steps.

In our laboratory we have concentrated our efforts on studies which are designed to provide information on the extent or complexity of RNA transcription from the repetitive and non-repetitive DNA base sequences found in the genomes of higher organisms.

The size of the mammalian genome is so large, compared to that of prokaryotes, that the number of DNA base sequences required to code for all of the known proteins is equivalent to a very small fraction of the total genomic information in any given cell. The mammalian genome contains an

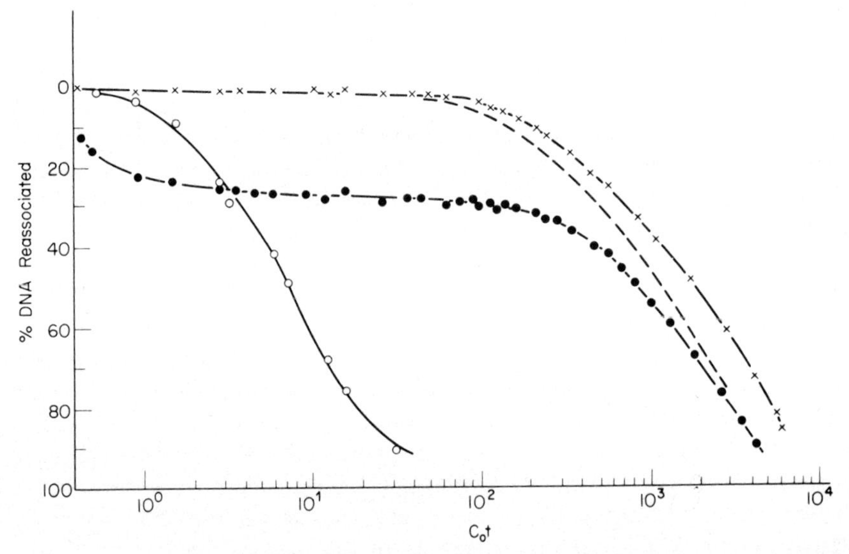


**Figure 1.** The reassociation kinetics of *E. coli* (○), total (●) and single copy (x) mouse DNA. Unlabeled DNA isolated from mouse liver was mixed with single copy $^3$H-DNA mouse (fractionated $C_0t$ 220) and $^{14}$C-*E. coli* DNA. The DNA mixture was sheared to a single stranded molecular weight of 120,000, heat denatured (100°C–10 minutes) and incubated in 0.12 M phosphate buffer at 60°C for various periods of time. 100 μg aliquots were fractionated on hydroxyapatite columns to determine the percentage of denatured DNA remaining at various times as described by Bell and Church (1971). The $C_0t$ values for single copy mouse and *E. coli* DNA are based on the concentration of unlabeled mouse DNA. The dashed line represents the theoretical second order reassociation kinetics expected for non-repeated DNA base sequence components in a reassociation reaction involving DNA isolated from a genome the size of that of the mouse (Britten, 1969).

enormous amount of genome complexity; genome complexity being defined as the number of distinct DNA base sequences present in the haploid sperm complement of a species (Laird, 1971). Genome complexity is determined by genome size, approximately 6 x $10^{-12}$g of DNA per cell in mammals, and by the relative repetition of base sequences in the genome. On the basis of DNA/DNA reassociation studies, Britten and Kohne (1968) established that the eukaryotic genome is composed of DNA base sequences which are found in vastly different frequencies ranging from non-repeated to highly repeated. Mammalian DNA base sequences can arbitrarily be assigned to three classes so that in the mouse; 6 to 10% of the genome is made up of very rapidly reassociating highly repeated DNA sequences; 25 to 35% reassociate faster than expected for a genome the size of that of the mouse indicating varying degrees of repetition, and the remaining 60-65% reassociates as expected for non-repeated single copy base sequences.

Figure I illustrates a typical reassociation profile for $^3$H labeled single copy DNA mouse sequences in a mixture containing a vast excess of unlabeled total mouse DNA. The reassociation profile of the total mouse DNA shows the "very rapidly" and "rapidly" renaturing fractions as well as the single component reassociation profile observed for the previously fractionated single copy DNA. The highly redundant sequences have not shown any evidence of being transcribed in any tissue or stage of development examined so far (Flamm *et al.*, 1969). *In situ* hybridization has shown these highly redundant sequences to be located in the centromere (Pardue and Gall, 1970) and in the heterochromatic regions of the chromosomes (Yasmiveb and Yunis, 1969). The very rapidly reassociating fraction of the mouse genome includes the AT satellite observed in cesium chloride gradients (Walker, 1969). Sequence analysis of this AT rich fraction suggests that the repeating base sequence units have a corrected complexity of approximately 140 base pairs (Sutton and McCallum, 1971) and occurs in a million copies since reassociation is $10^6$ times faster than would be expected if each sequence were present only once per mouse genome (Britten and Kohne, 1968).

The repeated or redundant fraction of the genome which reassociates at an average from a few to 100,000 times faster than expected (see Fig. 1) for non-repeated sequences in a genome the size of that of the mouse is that fraction which is the major DNA component in low $C_0t$ DNA/RNA hybridization studies (Church and McCarthy, 1968; McCarthy and Church, 1970). Under normal conditions the assay and interpretation of RNA transcriptional complexity from repeated DNA base sequences is limited (McCarthy and Church, 1970), since the chance of random collision between labeled RNA molecules and complementary nonrepeated DNA sequences fixed to a filter is most infrequent due to the genome complexity found in mammals (Britten, 1969). Many comparisons of RNA populations in different mammalian tissues have been reported in the last few years based on low $C_0t$ hybridization reactions without due regard to the criteria used in the reaction. The criteria is the specific base mismatching allowed by the RNA/DNA reassociation reaction conditions (McCarthy and Church, 1970). The presence of DNA sequences represented by families of similar but not identical DNA sequences of varying repetitive frequency means that RNA/DNA hydridization rarely if ever, displays cistron specificity (Church and McCarthy, 1968). Therefore, a single stranded DNA fragment and its complementary RNA transcript will seldom reassociate with each other since the probability of their reassociating with other members of the DNA base sequence family is greater (McCarthy and McConaughy, 1968). The reassociation criteria–salt concentration, concentration of organic solvents, temperature, nucleic acid fragment size and time–will determine the degree of RNA/DNA base sequence mismatching which will be permitted. The degree of sequence specificity is reflected in the melting profile and Tm of the reassociation duplexes. The choice of a particular reassociation criteria for an experimental system utilizing mammalian repeated DNA base sequences is an arbitrary one which allows minimum percentage of base pair mismatching, is adequate to analyse and is consistent with the specific activity of the labeled nucleic acids. The DNA-filter procedures usually used in these studies do not permit assay of RNA transcription from all frequency classes of DNA. In view of reactant conditions and incubation times it appears likely that most previous studies are limited to analysis of RNA transcribed from the repeated DNA sequences (McCarthy and Church, 1970).

The single copy $^3$H-DNA reassociation profile shown in Fig. 1 exhibits a rate of reassociation consistent with a single order reaction component with a reaction rate about 1,000 times slower than that of *E. coli* DNA. The total unlabeled mouse DNA sequences renatured at rates consistent with a multi-component system which includes a major single copy fraction equal to 65% of the genome. These experiments suggest that the purified single copy $^3$H-DNA sequences are essentially free of repetitive sequences and is, therefore, useful in the analysis of complexity of RNA transcription from single copy DNA.

The analysis of RNA sequences complementary to non-repeated DNA involves the reaction of DNA of high specific activity with high concentrations of unlabeled RNA for extended time periods (Kohne, 1968). This RNA excess reaction is in direct contrast to the DNA excess reaction developed by Melli *et al.*, (1971). Studies specifically designed to detect transcription of RNA complementary to non-repeated DNA have demonstrated that these sequences are transcribed in the mouse embryo (Gelderman *et al*, 1971; Church and Brown, 1972) and have been extended to an analysis of the transcriptional complexity in the *Xenopus* oocyte (Davidson and Hough, 1971). Estimates of the complexity of RNA transcription of single copy DNA sequences in various tissues of the adult mouse have been reported (Hahn and Laird, 1971; Brown and Church, 1971a, b; B. J. McCarthy, personal communication). The RNA/DNA reassociation product formed under high $C_0t$ reaction conditions has a melting profile and Tm which approaches the Tm of DNA/DNA reassociation duplexes (Brown and Church, 1971a, b).

I would like to report the results of our analysis of the complexity of RNA transcription complementary to non-repeated DNA at different stages of development of the mouse embryo and during specific stages of development of particular tissues (Brown and Church, 1971a, b). In these experiments high specific activity $^3$H-DNA was prepared from mouse L cells labeled with 3$\mu$c per/ml of $^3$H-thymidine. The purified DNA was sheared, heat denatured and then permitted to reassociate in 0.12 M phosphate buffer, pH 7, at 60°C for various periods of time. Aliquots of DNA were withdrawn at various time

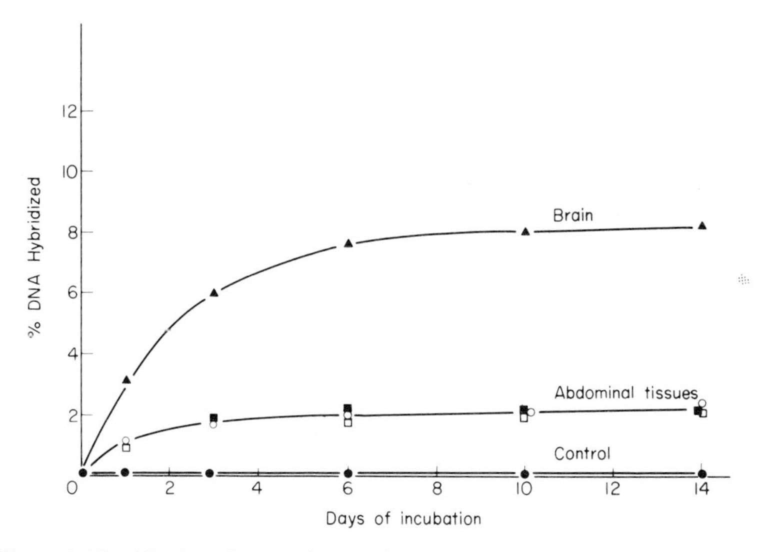


**Figure 2.** The kinetics of reassociation of RNA isolated from liver, kidney, spleen and brain of six week old mice in the reaction with single copy $^3$H-mouse DNA. The reassociation kinetics were derived from a reaction consisting of 0.5 $\mu$g of single copy $^3$H-DNA ($C_0t$ 220), 300 $\mu$g of brain (▲), liver (■), kidney (□) or spleen (○) RNA in 50 $\mu$l of 0.12 M phosphate buffer in sealed catillary tubes, heated to 100°C for 10 minutes and incubated at 60°C for the times indicated. The reaction was stopped by rapid cooling and the mixture either treated with RNAse or passed over a G200 column before loading onto a hydroxyapatite column. Single stranded nucleic acid was eluted with 0.12 M phosphate buffer at 60°C. The double stranded duplexes were either eluted from the hydroxyapatite with 0.5 M phosphate buffer, or melted from the hydroxyapatite column by stepwise temperature increments to 95°C. The percentage of single copy $^3$H-DNA hybridized to RNA was calculated after subtraction of the 0.5% hydroxyapatite column background obtained by alkaline or ribonucleased control (●). (Brown and Church, 1971b.)

periods, loaded onto hydroxyapatite columns equilibrated to 60°C with 0.03 M phosphate buffer. Double and single stranded high molecular weight DNA was bound to the column on loading and the single stranded DNA subsequently eluted by increasing the phosphate buffer concentration to 0.12 M. Double stranded DNA was eluted by 0.5 M phosphate buffer or by a stepwise increase in column temperature to 95°C while still in 0.12 M buffer. Optical reassociation was monitored at 260 nm in a continuously recording Gilford 2400 Spectrophotometer and the percent reassociation calculated by the $C_0t$ method of Britten and Kohne (1968). Isolation of single copy DNA base sequences was accomplished by reassociation to $C_0t$ 220 in 0.12 M phosphate buffer at 60°C as described previously (Brown and Church, 1971a). Total cellular RNA was isolated by a modified phenol method (Brown and Church, 1971a). All RNA was DNAsed extensively and brain RNA received CTA-bromide treatment. The RNA/non-repeated DNA reassociation reactions utilized sheared, heat denatured, non-repeated $^3$H-DNA in reaction with various purified RNA preparations (ratio >1/600) in 0.12 M phosphate buffer at 60°C. 50 $\mu$l Aliquots were sealed in glass capillary tubes and removed at timed intervals for analysis on hydroxyapatite. Control experiments indicated that RNA degradation during the reaction was minimal. When RNAse treated RNA preparations were incubated with non-repeated $^3$H-DNA less than 0.5% of the counts remained bound to the hydroxyapatite column in 0.12 M phosphate buffer at 60°C. All values were corrected to allow for this hydroxyapatite column background (Brown and Church, 1971a).

The kinetics of reassociation of non-repeated $^3$H-DNA with several different preparations of RNA isolated from different mouse tissues is presented in Fig. 2. The extent of reaction of single copy DNA with brain RNA is significantly greater than the reaction with RNA from the other tissues. The RNA/DNA duplexes formed cannot be attributed to traces of DNA contaminating the RNA preparation since the reaction is eliminated by RNase

**Table 1** Extent of RNA transcription from non-repeated Mouse DNA*

| Source of Unlabeled RNA<br>Average of three experiments | Percent Single copy $^3$H-DNA Reacted<br>Stage of Development<br>Newborn | <br><br>2 week | <br><br>6 week |
|---|---|---|---|
| Whole animal | 9.6 | – | 10.7 |
| Total brain | 5.2 | 8.5 | 8.2 |
| Liver | 2.5 | 4.1 | 2.0 |
| Kidney | 2.2 | 3.5 | 1.8 |
| Spleen | 2.0 | 3.0 | 2.0 |

* Methods used for RNA/DNA reassociation and hydroxyapatite analysis as described previously (Brown and Church, 1971b).

treatment prior to incubation and by the exclusion properties of the post incubation RNA from Sephadex-G 200.

A summary of the extent of reaction by $C_0t$ >8,000 of non-repeated $^3$H-DNA with RNA isolated from total mouse embryo and various tissues of the mouse is shown in Table 1. Approximately 10, 5 and 2% of the single copy $^3$H-DNA reacted with RNA isolated from whole newborn mouse, newborn brain, and newborn abdominal tissues respectively. The complexity of RNA transcription in brain tissue increases rapidly with development to approximately 8.5% by two weeks post partum, a value close to the complexity of transcription of the whole foetus at birth. In contrast the apparent saturation values of liver, kidney and spleen RNA isolated from six weeks old mice in reaction with single copy $^3$H-DNA were similar for all of the various stages of development examined. No differences in melting profiles or Tm were noted for the duplexes (Brown and Church, 1971a). Therefore, very little base pair mismatching exists in the duplexes formed with RNA isolated from different stages of development. The rate of reassociation of single copy $^3$H-DNA with RNA isolated from newborn brain RNA proceeds more slowly than the reaction with RNA isolated from six week old mouse brain (Brown and Church, 1971b).

The apparent saturation values found in these experiments probably reflects minimal complexity of RNA transcription in these mouse tissues. We have not as yet examined the complexity of RNA transcription from a single cell type by the methods outlined above. The fact that several cell types contribute to the results obtained for each tissue can be illustrated by additive experiments in which RNA from newborn and adult liver are simultaneously hybridized to single copy $^3$H-DNA. The apparent saturation value of these experiments was 3.5% compared to values of 2.5% found for newborn liver RNA and 2% for adult liver RNA. These results suggest that some RNA transcripts are common to newborn and adult stages but that others are unique to certain stages in tissue development. Complete additivity would suggest that RNA transcribed from single copy DNA at the two stages of development examined was comprised of two distinct RNA populations whereas a lack of additive nature in these additive experiments would suggest commonality. Additive experiments with liver and spleen RNA demonstrate partial overlap of complementary sequences, thus some RNA transcripts seem to be in common to both tissues while some are unique transcriptional products from each tissue (Brown and Church, 1971b).

In the present study the RNA driven RNA/DNA reassociation reaction was utilized to analyse patterns of transcriptional complexity complementary to the non-repeated single copy portion of the mouse genome during tissue development. In brain tissue the extent of RNA/single copy DNA reassociation increased rapidly during the first two weeks of life. The complexity of

RNA transcription in liver, kidney and spleen however appeared to remain fairly constant with the growth of the animal. In all of the developmental stages of the mouse examined, brain tissue had a greater complexity of RNA transcription from single copy DNA than did any other tissue examined. Hahn and Laird (1971) estimate the genetic complexity of RNA transcribed of single copy DNA in the adult mouse brain to be equivalent to more than 300,000 different DNA sequences, each 1000 nucleotides in length. This complexity reflects an apparent summation of the total gene activity of the numerous cell types represented by RNA isolated from total brain tissues. The analysis reported here represents total RNA transcripts present in a given tissue, which are complementary to single copy DNA, such that stable RNA transcripts synthesized at earlier stages in development could contribute to the total saturation value at a particular stage examined. Comparison of RNA isolated from different tissues indicates that some RNA transcripts were common while others were tissue specific. The significance of the high complexity of RNA transcription in total brain tissue cannot meaningfully be interpreted at this time. Further experiments utilizing homogeneous cell populations will be required to determine the contribution that neuronal, glial and other cell types make to the extent of genetic complexity noted in these experiments.

## ACKNOWLEDGEMENTS

The authors are grateful for the excellent technical assistance provided by Mrs Judy Crozier and Mrs Anne Vipond. They also wish to acknowledge the assistance of Mr G. Bell in the hydroxyapatite chromatography. This work was supported by a Medical Research Council of Canada Fellowship (I.B.) and operating grants from the National Research Council and National Cancer Institute of Canada.

## REFERENCES

Bell, G. I. and Church, R. B. (1972). *Biochem. Genet.* in press.

Britten, R. J. (1969). *In* Problems in Biology: RNA in Development (ed. W. E. Hanly), pp. 187-210, University of Utah Press.

Britten, R. J. and Kohne, D. E. (1968). *Science, N.Y.* **161**, 529-540.

Brown, I. R. and Church, R. B. (1971a). *Biochem. biophys. Res. Commun.* **42**, 850-856.

Brown, I. R. and Church, R. B. (1971b). *Devl Biol.* in press.

Church, R. B. and Brown, I. R. (1971). *In* Results and Problems in Differentiation, Vol. 3. (ed. H. Ursprung), Springer-Verlag. pp. 11-24.

Church, R. B. and McCarthy, B. J. (1968). *Biochem. Genet.* **2**, 55-87.

Davidson, E. H. and Hough, B. (1971). *J. molec. Biol.* **56**, 491-506.

Flamm, W. G., Walker, P. M. B. and McCallum, M. (1969). *J. molec. Biol.* **40**, 423-443.

Gelderman, A. H., Rake, A. V. and Britten, R. J. (1971). *Proc. natn. Acad. Sci. U.S.A.* **68**, 172-176.
Hahn, W. E. and Laird, C. D. (1971). *Science, N.Y.* **173**, 158-161.
Harris, H. (1963). *Prog. Nucl. Acid. Res.* **2**, 19-38.
Kohne, D. E. *(1968). Biophys. J.* **8**, 1104-1112.
Laird, C. D. (1971). *Chromosoma* **32**, 378-394.
McCarthy, B. J. and Church, R. B. (1970). *A. Rev. Biochem.* **39**, 131-150.
McCarthy, B. J. and McConaughy, B. L. (1968). *Biochem. Genet.* **2**, 37-53.
McCarthy, B. J., Shearer, R. W. and Church, R. B. (1969). *In* Problems in Biology: RNA in development. (ed. W. E. Hanly), pp. 285-312, University of Utah Press.
Melli, M., Whitfield, C., Rao, K., Richardson, M. and Bishop, J. O. (1971). *Nature Lond.* **231**, 8-12.
Pardue, M. L. and Gall, J. D. (1970). *Science, N.Y.* **168**, 1356-1358.
Sutton, W. D. and McCallum, M. (1971). *Nature Lond.* **232**, 83-84.
Walker, P. M. B. (1969). *Prog. Nucl. Acid Res.* **9**, 301-342.
Yasmiveb, W. and Yunis, J. (1969). *Biochem. Biophys. Res. Comm.* **35**, 779-784.

FEBS Symposium, Volume 24, 1972, pp. 109-111

# Inhibition of Erythropoiesis by Bromodeoxyuridine

P. MALPOIX and B. DOEHARD

*Laboratoire de Cytologie et d'Embryologie Moléculaire*
*Université Libre de Bruxelles, Bruxelles, Belgium*

Bromodeoxyuridine has been shown to specifically inhibit the synthesis of the terminal differentiated proteins in myogenic cells, in which the mitotic and post-mitotic phases of differentiation are clearly distinguished in time. During erythropoiesis, the synthesis of hemoglobin proceeds in cells which are still actively synthesizing DNA and dividing. Miura and Wilt (1971) have already shown that hemoglobin synthesis can be inhibited at early stages of yolk sac erythropoiesis in chick embryos. Holtzer, in this symposium, has situated this inhibitory effect in hematocytoblasts. Does BUdR exert a uniquely specific effect on hemoglobin formation in such cells, or are concomitantly synthesized chromatin proteins also sensitive to this inhibitor? In other words, does BUdR incorporation into DNA inhibit an initiation step or a trigger mechanism specifically requisite for certain kinds of tissue specific proteins, or does it also diminish the synthesis of other proteins of universal importance in all cells for the maintenance of basic structure and function.

To answer this question, we have used a sensitive method, previously described in detail (Malpoix *et al.*, 1969; Malpoix, 1971), to detect differential effects on the synthesis of proteins, including Hb, at different stages of erythropoiesis in disaggregated fetal mouse liver cultivated *in vitro*. Erythropoietin stimulates differentiation in this tissue, acting selectively on the most immature erythroid precursors, but exerting multiple effects on DNA, RNA, chromatin protein and hemoglobin synthesis (Goldwasser, 1966; Krantz and Jacobson, 1970; Malpoix, 1971; Marks and Rifkind, 1972). The potential inhibitory action of bromodeoxyuridine has therefore also been examined in relationship to erythropoietin action.

The effective penetration of $^{3}$H-bromodeoxyuridine into the cultured cells has been checked by autoradiography and by ultra-centrifugation in CsCl gradients. Nuclei are heavily labeled as early as 1 h after incubation in media containing the labeled molecule. The degree of increase in buoyant density of newly synthesized DNA depends on the concentration of the inhibitor: after

20 h, heavy peaks were observed with buoyant densities of 1.753, 1.751, 1.746 and 1.740 with 100, 50, 10 and 5 $\mu$g BUdR/ml respectively. In comparison, controls only displayed the usual major peak at 1701 and the mouse satellite at 1691. Interestingly, erythropoietin, which stimulates the synthesis of DNA, significantly increases the incorporation of labeled bromodeoxyuridine. Another still unexplained, but interesting finding, is an early and transient stimulation of $^3$H-guanosine incorporation into DNA in the presence of low concentrations of BUdR (5 and 10 $\mu$g/ml).

**Table 1.** Effects of BUdR ($3.10^{-5}$ M) on normal and stimulated erythropoiesis in disaggregated fetal mouse liver cultured *in vitro*.

| | Specific radioactivity expressed in cpm/$\mu$mol $^3$H-Lysine | | | | | |
|---|---|---|---|---|---|---|
| | Nucleus | | | | Cytoplasm | |
| | Histones | | Non-Histones | | Hemoglobin | |
| Controls[1] | 31,050 | | 30,226 | | 34,586 | |
| Treated erythropoietin[2] | 54,502 | +75.5% | 42,566 | +40.2% | 72,978 | +111% |
| Treated erythropoietin[2] + bromodeoxyuridine[3] | 19,456 | −37.3% | 15,458 | −48.8% | 3,323 | −90.4% |
| Treated bromodeoxyuridine[4] | 10,600 | −65.9% | 7,875 | −73.9% | 2,216 | −93.6% |

[1] Untreated for the whole incubation period, a total of 46 h.
[2] Erythropoietin added after 20 h
[3] Pretreatment for 20 h with BUdR before addition of EP.
[4] BUdR added at the beginning of incubation.

$^3$H-Lysine added at 40 h for the final 6 h in all cases

Inhibition of the synthesis of hemoglobin is highest when young cultures containing a high percentage of hematocytoblasts and proerythroblasts are treated: for example, in one experiment *in vitro* cultures of livers derived from 11.5 day embryos responded to a pretreatment of 18 h with 10 $\mu$g BUdR/ml followed by 5 h continued treatment in the presence of $^3$H-Lysine, by a fall of 71% in Hb synthesis. Nuclear proteins, in the same experiment, were only inhibited by 43%. In another experiment, in which effects on histones and acidic proteins were distinguished, the synthesis of histones was again found to be less inhibited than hemoglobin, while non-histone synthesis remained unaffected. Inhibitions of Hb synthesis of up to 80% and 90% were observed in some cases, when cultures of extremely immature cells were used.

Later stages of erythropoiesis were clearly much less sensitive, however, since similar concentrations of BUdR used to treat disaggregated liver from 14 and 15 day embryos, in which most of the erythroid cells had proceeded to later stages of maturation, inhibited Hb synthesis by 20% or less, and had little effect on histone or acidic protein synthesis.

Most significant, perhaps, is the general finding that in actively dividing and differentiating cells, the inhibitory effect of bromodeoxyuridine is *not* an all or none reaction affecting only hemoglobin synthesis: nuclear proteins are also sensitive, though to a lesser extent. This feature of BUdR action is particularly clear when examined in relationship to erythropoietin (EP) action in cultures derived from 11 and 12 day embryos (Table 1). The stimulatory effect of EP on both chromatin proteins and hemoglobin is almost eliminated by BUdr provided the inhibitor is added 20 h before the hormone.

Thus, bromodeoxyuridine seems able to inhibit the whole complex process of erythropoiesis when the inhibitor is effectively incorporated into DNA during the early stages of differentiation. The fact that the synthesis of chromatin proteins is also affected by the inhibitor has implications with respect to the mechanism of action of BUdR, which still remains unexplained.

## REFERENCES

Goldwasser, E. (1966). *In* Current Topics in Developmental Biology, (A. Monroy and A. A. Moscona, eds.) Academic Press, New York and London.

Krantz, S. B. and Jacobson, L. O. (1970). *In* Erythropoietin and Regulation of Erythropoiesis, The University of Chicago Press, Chicago.

Malpoix, P. (1971). Expl. Cell Res. 65, 393.

Malpoix, P., Zampetti, F. and Fievez, M. (1969). *Biochim. biophys. Acta* **182**, 214.

Marks, P. and Rifkind, R. (1972). *Science, N.Y.* (in press).

Miura, Y. and Wilt, F. H. (1971). *J. Cell. Biol.* **48**, 523.

FEBS Symposium, Volume 24, 1972, pp. 113-116

# Myosin Synthesis During Morphological Differentiation of Myoblast Lines

D. LUZZATI, G. DRUGEON and W. F. LOOMIS, Jr*

*Institut de Biologie Moléculaire Faculté des Sciences, Paris, France*

The isolation of established lines of myoblasts, from rat neonatal skeletal muscle, by Yaffé in 1968 has given the developmental biologist an extremely interesting biological system for the study of the terminal steps in cell differentiation. Myoblast cell lines, such as line L 6 used in this study, can be maintained indefinitely in an undifferentiated (but predetermined) state by trypsinizing the culture before the onset of differentiation. If incubation is pursued after confluency (c.a. $2 \times 10^6$ cells/60 mm Falcon plate) the cells start to fuse into multinucleated myotubes and to synthesize actively characteristic muscle proteins, either structural, like myosin and actin (Luzzati and Drugeon, in preparation), or enzymatic, like phosphorylase, phosphocreatine kinase or myokinase (Shainberg *et al.*, 1971; Wahrmann and Luzzati, in preparation).

In order to elucidate the number of steps involved in the transition from predetermined to committed state in this system, as a prerequisite to unravelling the underlying molecular mechanisms, we have isolated thermosensitive developmental mutants of line L 6 (Loomis and Luzzati, in press). Four classes of mutants have been found: class I and class II mutants do not grow at high temperature (41°C) and while class I strains differentiate at 37°C, class II mutants do not fuse into myotubes at low temperature. Class III and class IV mutants both grow at 37°C and 41°C at the same rate as the wild type and while class III mutants do not differentiate at high temperature, class IV mutants differentiate at 41°C, but not at 37°C.

We have used two of these mutants, E 3 from class III, and H 6 from class IV, to investigate the temporal and causal relationship between morphological and biochemical differentiation. Morphological differentiation has been evaluated by daily microscopic observation of the cultures and estimation of the number of myotubes per surface area of the plates while biochemical

* Present address: Department of Biology, University of California, San Diego La Jolla, U.S.A.

differentiation has been followed by measuring the rate of synthesis and the accumulation of muscle major protein, myosin. The results reported here are still rather preliminary, for myosin is not a very easy protein to handle. However, we have evolved an SDS polyacrylamide method which allows reasonably reliable determination of labeled as well as unlabeled myosin concentration in cytoplasmic cell extracts (Luzzati and Drugeon, in preparation) (Fig. 1).

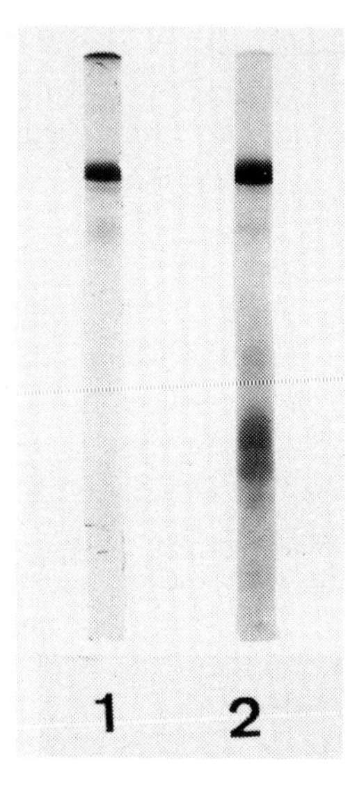


**Figure 1.** Disc electrophoresis of muscle contractile proteins. Polyacrylamide SDS gels of (1) purified myosin; (2) extract of differentiated cells of line $L_6$.

Exponentially growing undifferentiated myoblasts already contain low levels of myosin (c.a. 1% of total cell's proteins), and they incorporate radioactive amino acids into myosin at a low rate. At the beginning of morphological differentiation the relative rate of synthesis of myosin increases abruptly and proceeds at a high rate during the whole fusion period (Fig. 2) leading to an important accumulation of this structural protein (up to 40% of cytoplasmic proteins). High temperature sensitive developmental mutant E 3 exhibits an enhanced rate of myosin synthesis at the permissive temperature, more or less parallel to the increase in myotube formation, but the rate of synthesis does not increase at 41°C, where very little fusion is observed. The reverse pattern is observed with cold sensitive developmental mutant H 6, which neither fuses nor displays a significant increase in the rate of myosin synthesis at 37°C but differentiates, morphologically and to a slight extent biochemically, at 41°C.

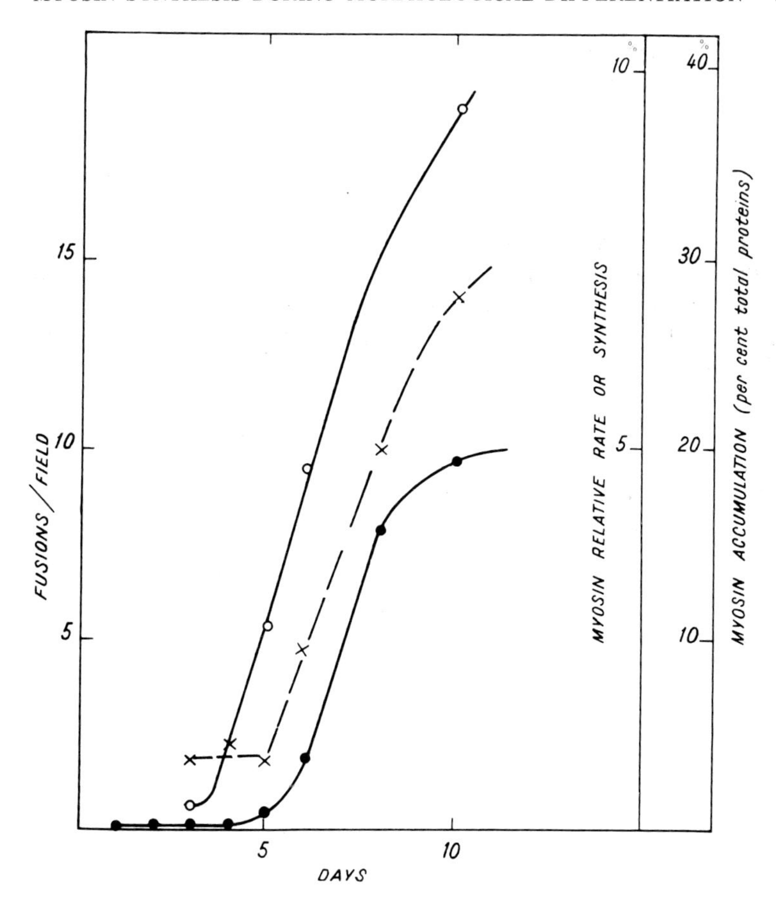


**Figure 2.** Morphological and biochemical differentiation of myoblast line $L_6$ – · – number of fusions per unit area of culture plate. (100 areas were examined at random for each experimental point). – x . . . x . . . myosin rate of synthesis (percentage of radioactive precursor incorporated into myosin during a labeling period of 3 h, with respect to radioactivity incorporated in total cytoplasmic proteins). – x – x – myosin accumulation ($\mu$g of myosin % $\mu$g of total cell cytoplasmic proteins).

These experiments show that there is a definite correlation between morphological and biochemical differentiation in this system; they strongly support the conclusion that both fusion process and active synthesis of specific muscle proteins might be triggered by the same developmental event.

Temperature shift experiments with the mutants (Loomis and Luzzati, unpublished) as well as medium shifts (Yaffé, 1968) point out that this event or series of events, occur during a short and well defined period, possibly after a decisive kind of mitosis (Holzer, this symposium), which precedes the beginning of a visible fusion process. Nothing is yet known about the nature of the triggering event (or events) although many hypotheses have already been put forward ranging from changes at the DNA level such as gene amplification (Crippa and Tocchini-Valentini, this symposium, Bishop and Pemberton, ibid, Luzzati, unpublished), changes in DNA transcriptional patterns through either RNA polymerase alterations or modifications of regulatory proteins, to changes in translational control mechanisms (Heywood, 1970). Experiments are under way in our, as well as in other, laboratories to test these hypotheses. It may already be expected that changes in more than one type of control mechanisms are involved in the transition from predetermined to differentiated state in order to insure the stability of the latter.

## REFERENCES

Heywood, S. (1970). *Proc. natn. Acad. Sci. U.S.A.* **67**, 1782.
Shainberg, A., Yagil, G. and Yaffé, D. (1971). *Dev. Biol.* **85**, 1.
Yaffé, D. (1968). *Proc. natn. Acad. Sci. U.S.A.* **61**, 477.

FEBS Symposium, Volume 24, 1972, p. 117

# Synthesis of a Collagen-like Protein in Sea Urchin Embryos

IDA PUCCI-MINAFRA

*Institute of Comparative Anatomy*
*The University of Palermo, Italy.*

When sea urchin embryos are incubated in the presence of $^{3}$H-Proline ($^{3}$H-Pro) part of the radioactivity incorporated into proteins is recovered as $^{3}$H-Hydroxyproline ($^{3}$H-HyPro).

In the cleavage stages, the $^{3}$H-HyPro/$^{3}$H-Pro ratio is quite low (0.3%); it undergoes a 15 fold increase at the gastrula stage. This increase is concurrent with the appearance of the spicules. At the pluteus stage the $^{3}$H-HyPro-containing protein is almost exclusively recovered from the organic matrix of the spicules.

From the characteristics of extractability of the $^{3}$H-HyPro-containing-protein and from its high $^{3}$H-HyPro/$^{3}$H Pro ratio (20%) it can be suggested that it is either collagen or a collagen-like protein.

It could be further suggested that this protein plays a role in the differentiation of the spicules.

FEBS Symposium, Volume 24, 1972, pp. 119-130

# Specific Biological Inhibitors of Protein Synthesis in Differentiated Cells

J. KRUH, F. LEVY and L. TICHONICKY

*Institut de Pathologie Moléculaire*
*Paris* 14e *France**

When a sample of pH 5 fraction prepared from liver post ribosomal fraction was added to a reticulocyte cell free system, it inhibited hemoglobin synthesis (Kruh and Levy, 1967; Levy *et al.,* 1970). Conversely, when a sample of reticulocyte pH 5 fraction was added to a liver cell free system, it inhibited serum albumin synthesis (Levy *et al.*, 1971) (Fig. 1). The addition of an extra amount of reticulocyte pH 5 fraction had no significant effect on hemoglobin synthesis, the addition of an extra amount of liver pH 5 fraction to liver cell free system slightly stimulated serum albumin synthesis. It can be concluded that the pH 5 fractions from these differentiated cells contain an inhibitor of protein synthesis in heterologous cells. It is probably a general phenomenon, since kidney, heart and spleen pH 5 fractions inhibit cell free synthesis of hemoglobin.

## ISOLATION OF THE INHIBITORS

These inhibitors are proteins, located in cell cytoplasm. Their purification requires two consecutive chromatographies: CM-cellulose which retains all non acidic proteins and DEAE-cellulose. The reticulocyte inhibitor was eluted from the second column by Tris buffer (pH 7.2) with a concentration gradient. Four protein peaks were obtained, the inhibitor was eluted with 0.03 M Tris buffer (Table 1). In the next step we used 0.02 M and 0.03 M Tris buffer.

* Groupe de l'Institut National de la Santé et de la Recherche Médicale, Laboratoire associé au Centre National de la Recherche Scientifique.

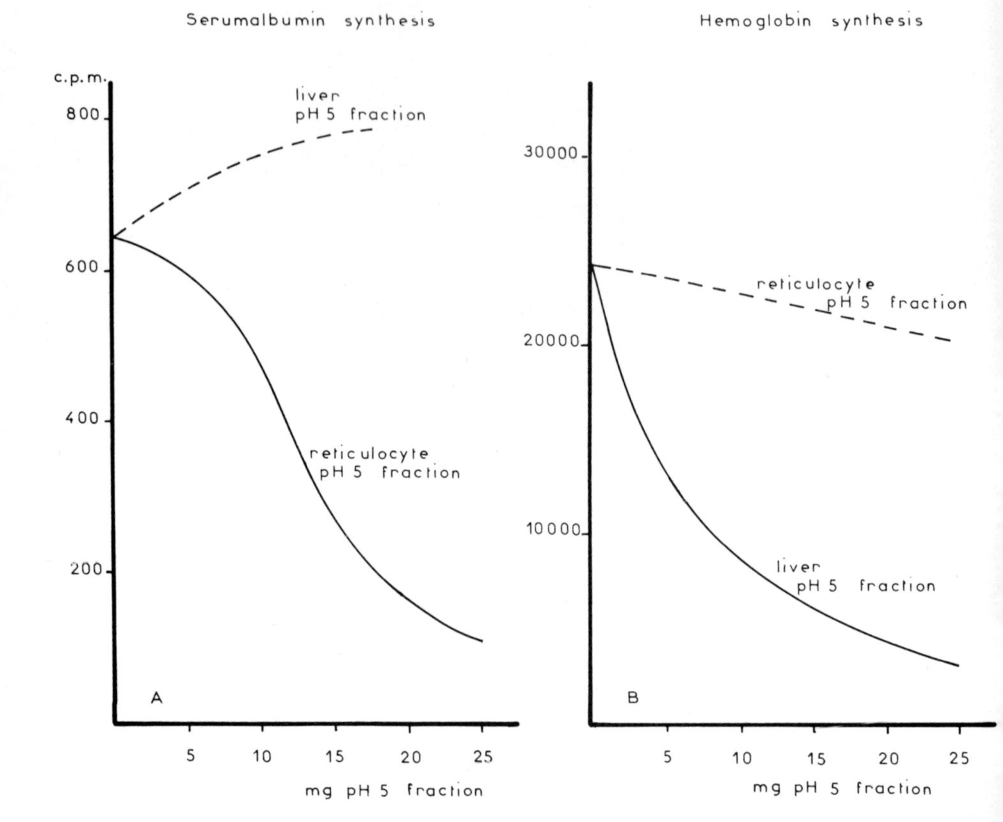


**Figure 1.** Action of pH 5 fractions on specific protein syntheses. A. Liver cell free system was made of (Von der Decken and Campbell, 1962; Levy, *et al.*, 1971): 0.5 ml of microsomal suspension (15 mg of protein), 0,3 ml of liver pH 5 fraction (12 mg of protein), 2 $\mu$moles of ATP, 25 $\mu$moles of creatine phosphate, 300 $\mu$g of creatine kinase, 0.25 $\mu$moles of GTP, 6 $\mu$moles of reduced glutathion and 2 $\mu$Ci of $^{14}C$ leucine (specific activity 40 mCi/mM). The final $Mg^{2+}$ concentration was 5 mM the final $K^+$ concentration, 75 mM. The system was incubated for 60 min at 37°C. 10 mg of serum albumin were added as a carrier. Sodium dodecylsulfate (SDS) was added to a final concentration of 1%. The proteins were precipitated by 1 volume of 10% trichloracetic acid. Serum albumin was extracted from the precipitate by 96% ethanol containing 1% trichloracetic acid. A dialysis against water precipitated the globulins still present. The purity of serum albumin was checked by cellulose electrophoresis. The radioactivity of the serum albumin was measured in a Nuclear Chicago Liquid Scintillator. The amount of the liver pH 5 fraction used here is suboptimal, the optimal amount is 18 mg of protein. B. Reticulocyte cell free system was made of (Schweet *et al.*, 1958; Kruh, 1968): 0.5 ml of ribosomal suspension (6 mg of protein) 0.5 ml of pH 5 fraction (10 mg of protein), 2 $\mu$moles of ATP, 25 $\mu$moles of creatine phosphate, 300 $\mu$g of creatine kinase, 0.25 $\mu$moles of GTP, 0.5 $\mu$Ci of $^{14}C$ leucine (specific activity: 40 mCi/mM). The final $Mg^{2+}$ concentration was 4.5 mM, the final $K^+$ concentration was 75 mM. 35 mg of hemoglobin were added as a carrier. Hemoblobin was isolated by starch block electrophoresis, its radioactivity was measured in a Mesco gas flow counter.

**Table 1**

| | Serum albumin radioactivity counts/min. | | | | |
|---|---|---|---|---|---|
| | | Cytoplasmic fraction added | | | |
| μg protein | Control | 0.03 M | 0.05 M | 0.07 M | 0.10 M |
| 200 | 650 | 253 | 426 | | 755 |
| 200 | 930 | 130 | 552 | | 1,050 |
| 250 | 1,125 | 285 | 510 | 960 | 1,600 |
| 375 | 625 | 65 | 242 | 640 | 835 |
| 500 | 900 | 70 | 200 | 800 | |

Synthesis of serum albumin by liver cell-free system in the presence of the various fractions obtained from the chromatography column with a discontinuous concentration gradient of Tris buffer (pH 7.2). The specific activity of $^{14}C$ leucine was 40 mCi/mM.

With the first elutant we obtained two peaks, with the second, one peak (Fig. 2). The inhibitor was present in peak I eluted with 0.02 M buffer (Table 2). A similar procedure was used with the liver pH 5 fraction, the inhibitor was eluted with 0.03 M Tris buffer (pH 7.2).

This procedure allowed a good purification of both inhibitors, the reticulocyte inhibitor was obtained in a more purified state. A polyacrylamide gel showed only one major peak (Fig. 3). The isoelectric point was measured by the electrofocusing technique, the migration of the protein corresponded to an isoelectric point of 5.4 (Fig. 4). The amino acid compositions of both inhibitors showed high amounts of glutamic and aspartic acids (Table 3). The molecular weight was estimated by the polyacrylamide–SDS technique which showed two bands of molecular weights 60,000 and 65,000. This could reflect either an heterogeneity or the presence of two subunits (Fig. 5).

**Table 2**

| | Serum albumin radioactivity counts/min. | | | |
|---|---|---|---|---|
| | | Cytoplasmic fraction added | | |
| | | 0.02 M | | |
| μg protein | Control | I | II | 0.03 M |
| 100 | 8,400 | 3,470 | 6,800 | 6,650 |
| 200 | 5,670 | 2,420 | 4.930 | 4,630 |

Synthesis of serum albumin by liver cell free systems in the presence of each of the two peaks obtained with 0.02 M Tris buffer and the peak obtained with 0.03 M Tris buffer (pH 7.2). The specific activity of $^{14}C$ leucine was 130 mCi/mM.

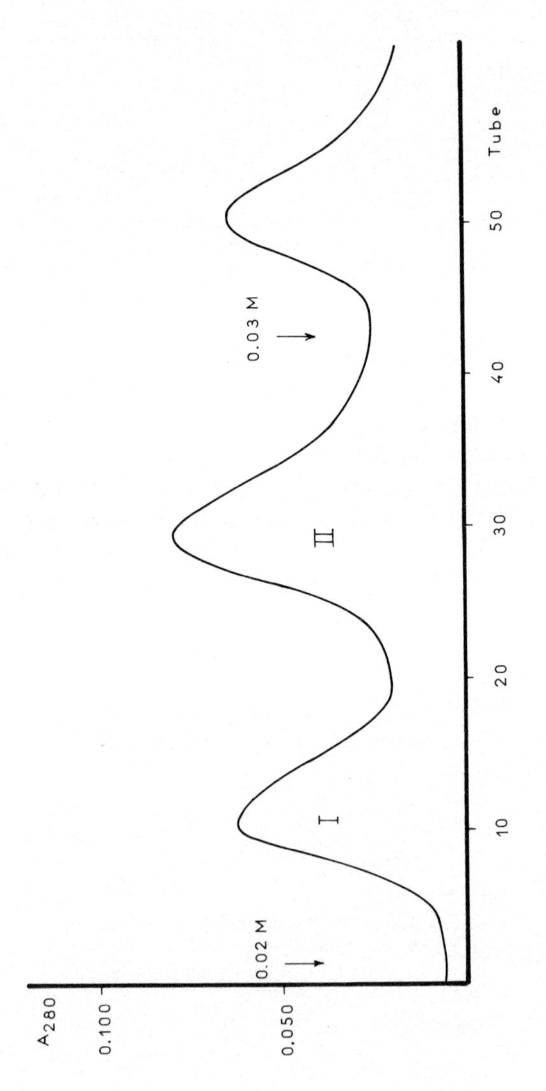


**Figure 2.** Preparation of the reticulocyte inhibitor. Reticulocyte pH 5 fraction, containing 300 mg of protein was transferred to a CM-cellulose column, eluted with 0.01 M Tris buffer (pH 7.2) directly on a DEAE cellulose column. This column was washed with the same buffer and the proteins were eluted with 0.02 M and 0.03 M Tris buffer (pH 7.2). Each peak contained 600-1000 $\mu g$ of protein.

**Figure 3.** Electrophoresis of reticulocyte inhibitor on polyacrylamide gel. Electrophoresis was performed according to Labrie (1969) with 5% acrylamide and 0.04 M tris buffer (pH 7.8) containing 0.02 M sodium acetate and 2 mM EDTA for 45 min at 5 mA. The protein was stained with Amido-Schwartz.

**Table 3.** Aminoacid composition of the inhibitors

| Aminoacid | Reticulocyte inhibitor | Liver inhibitor |
|---|---|---|
| Lysine | 4.1 | 6.1 |
| Histidine | 1.8 | 1.8 |
| Arginine | 2.4 | 3.5 |
| Aspartic acid | 10.2 | 8.6 |
| Threonine | 3.6 | 6.0 |
| Serine | 6.1 | 7.0 |
| Glutamic acid | 8.5 | 10.2 |
| Proline | 4.0 | 5.8 |
| Glycine | 6.1 | 8.2 |
| Alanine | 5.6 | 7.5 |
| Valine | 4.6 | 5.4 |
| Isoleucine | 2.7 | 3.8 |
| Leucine | 9.9 | 7.5 |
| Tyrosine | 2.2 | 3.1 |
| Phenylalanine | 3.0 | 3.4 |

The inhibitors were hydrolysed under vacuum for 22 h at 110°C with 1 ml of 6 N HCl containing 2 $\mu$M phenol per ml of protein. The aminoacid determination was performed with the automatic Beckman aminoacid analyser 120 C.

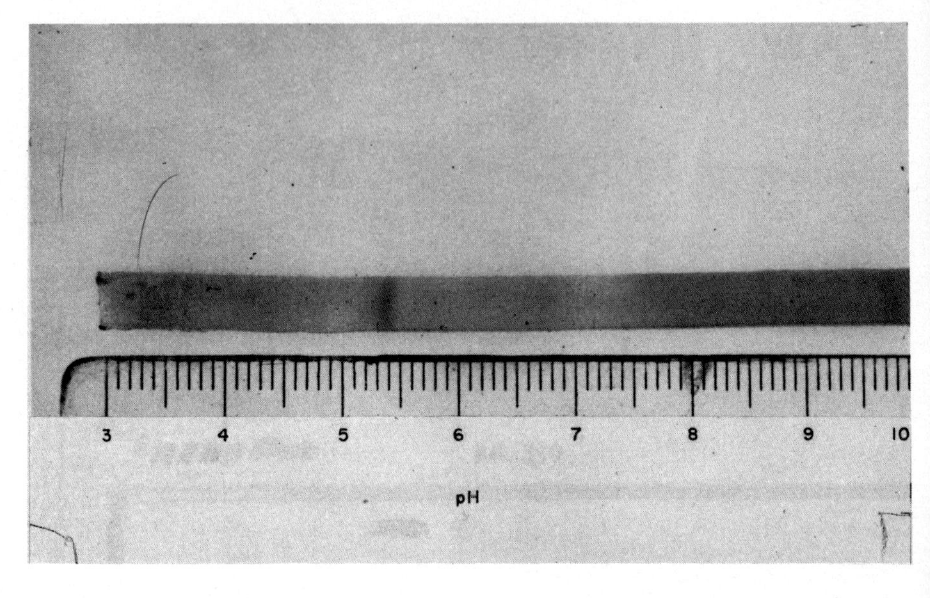


**Figure 4.** Electrofocusing of reticulocyte inhibitor. Electrofocusing was performed according to the procedure of Awdeh *et al.* (1968) modified by Wrighley (1968) with the use of LKB ampholine carrier with a pH gradient between 3 and 10. The protein was stained with Bromophenol Blue.

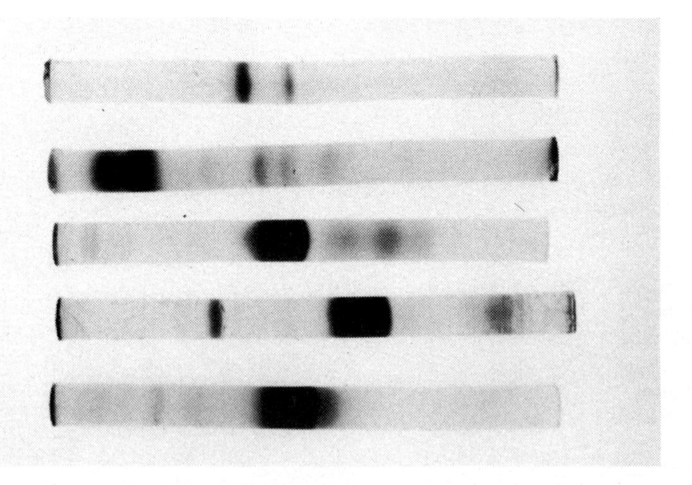

**Figure 5.** Molecular weight of reticulocyte inhibitor. The molecular weight was estimated by electrophoresis on polyacrylamide SDS gel for 15 h at 3 mA with 0.1 M sodium phosphate buffer (pH 7.2) (Schapiro *et al.*, 1967; Weber and Osborn, 1969). Starting from the top: Reticulocyte inhibitor, urease (83,000) phosphopyruvate kinase (57,000), phosphogluceraldehyde dehydrogenase 72,000-36,000), serum albumin (65,000).

## SPECIFICITY

The inhibitors have retained the specificity of the pH 5 fractions, they are very active on heterologous systems and inactive on homologous systems (Table 4).

**Table 4.** Specificity of the inhibitors

| Origin of the Inhibitor added | Radioactivity Counts/min. Serum albumin (liver cell free system) 1 | 2 | Hemoglobin (reticulocyte cell free system) 1 | 2 |
|---|---|---|---|---|
| Control | 1,800 | 1,625 | 37,000 | 44,000 |
| Liver | 2,920 | 1,310 | 5,450 | 5,500 |
| Reticulocyte | 418 | 252 | 34,200 | 45,000 |

Inhibitors extracted from liver and from reticulocytes and containing 200 $\mu$g of protein each were added to both liver and reticulocyte cell-free systems. The liver cell-free system was made of a microsomal suspension (15 mg of protein) and of pH 5 fraction (18 mg of protein), the $Mg^{2+}$ concentration was 5 mM. The reticulocyte cell-free system was made of a ribosomal suspension (6. mg of protein) and of pH 5 fraction (10 mg of protein), the $Mg^{2+}$ concentration was 4.5 mM. In all experiments the final volume was 1.5 ml. The specific activity of $^{14}C$ leucine was 40 mCi/mM.

## MECHANISM OF ACTION

These inhibitors are involved in protein synthesis in an highly specific step. We have checked that they did not interfere with the initial and final steps of protein formation: synthesis of aminoacyl-tRNAs, release and assembly of peptide chains. They probably do not act on elongation factors which are not tissue specific. The most likely mechanism would be an action on messenger RNAs.

RNA prepared from liver nuclei was incubated with the reticulocyte and with the liver inhibitors and subjected to centrifugation on a sucrose gradient. None of these factors acted on the 28S RNA. The reticulocyte inhibitor highly disturbed the 18S RNA and the RNA fraction which sedimented between 18S and 4S and which contained most of the messenger RNAs (Kruh *et al.*, 1966). This fraction was called RNA III. The liver inhibitor modified only slightly the behavior of RNA III on the gradient (Fig. 6). The inhibitors could react with RNAs either through a nucleolytic effect or by binding to some of them. We have tested unsuccessfully the nucleolytic hypothesis on several types of RNA: TYMV RNA, tRNA, polysomes (liver inhibitor on reticulocyte polysomes) and yeast RNA. A more specific experiment has been performed. We

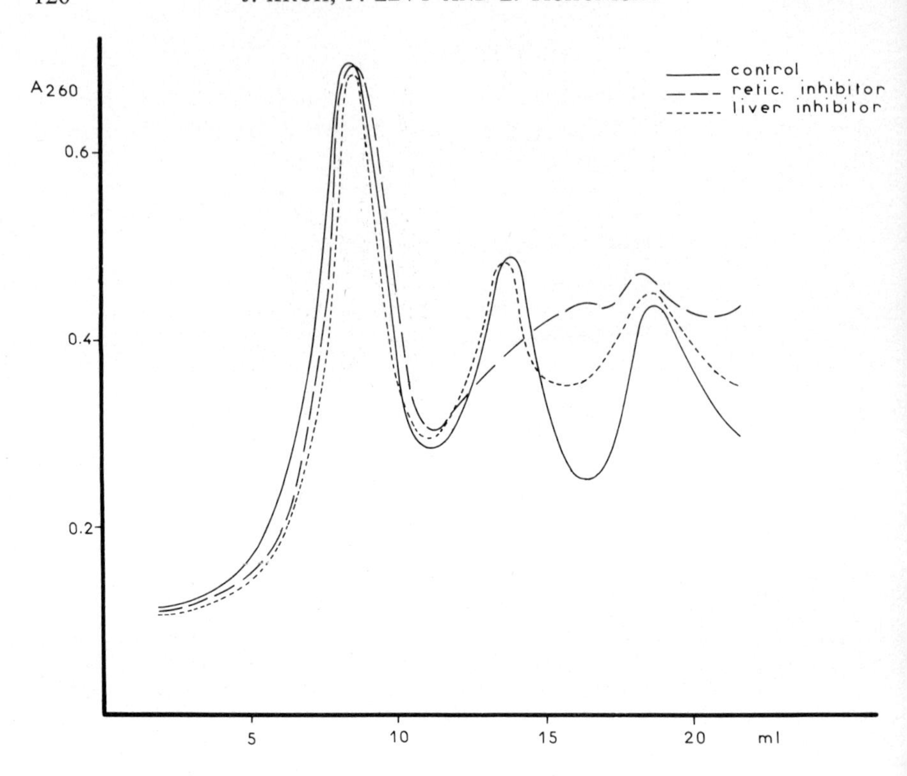


**Figure 6.** Action of reticulocyte and liver inhibitors on liver RNA on a sucrose gradient. 700 μg of liver RNA were incubated with 400 μg of either liver or reticulocyte inhibitor for 15 min at 4°C and transferred on a 5-20% sucrose linear gradient. The centrifugation was performed in a SW 25 Spinco Rotor for 17 h at 35,000 x g. RNA was prepared from liver nuclei using the technique of Hiatt (1962) slightly modified (Kruh, 1967).

have previously demonstrated that reticulocyte RNA is able to stimulate hemoglobin synthesis in a reticulocyte cell free system (Kruh *et al.*, 1964a), and that liver inhibitor suppresses this stimulation (Levy *et al.*, 1970). Reticulocyte RNA was incubated with the liver inhibitor and purified in order to eliminate the inhibitor. When this RNA was added to a reticulocyte cell-free system, it stimulated hemoglobin synthesis as well as non-pre-incubated RNA. It would not have been the case if the RNA had been split by the inhibitor (Table 5).

A binding of the inhibitors to RNAs has to be considered. The existence of such a binding was demonstrated as follows: Radioactive liver RNA was incubated with the inhibitor, the inhibitor was precipitated by lowering the pH to 5 (in this condition free RNA was not precipitated). The mixture was transferred to a millipore filter. The existence and the extent of the binding was deduced from the radioactivity retained on the filter. Such experiment was performed in two variants. In the first experiment, radioactive liver RNA was fractionated on a sucrose gradient into four fractions, including RNA III, and each of the fractions was incubated with the reticulocyte inhibitors. The

**Table 5.** Action of the inhibitor from liver and of reticulocyte RNA on hemoglobin synthesis by a reticulocyte cell-free system

| Incubation Medium | Hemoglobin Synthesis c.p.m. |
|---|---|
| Control | 18,000 |
| + 400 μg reticulocyte RNA | 70,000 |
| + 200 μg liver inhibitor | 3,010 |
| + 400 μg reticulocyte RNA +200 μg liver inhibitor | 2,950 |
| + 400 μg reticulocyte RNA preincubated with liver inhibitor and purified | 66,000 |

A recticulocyte cell-free system (Schweet *et al.*, 1958; Kruh, 1968) was incubated in the presence of $^{14}C$ leucine in the presence of either reticulocyte RNA prepared by the procedure of Kirby (1956) or liver inhibitor, or both of them. In the last incubation RNA was incubated for 15 min at 37° C with liver inhibitor, purified again by the Kirby technique and added to the cell free system. The incubation was carried out for 60 min at 37° C. Hemoglobin was purified on starch block electrophoresis.

**Table 6.** Binding of the liver RNA fractions obtained from a sucrose gradient to reticulocyte inhibitor

| Nature of $^3H$ RNA | Radioactivity of 100 μg RNA c.p.m. | Radioactivity of the RNA-inhibitor complex c.p.m. | Amount of RNA bound to 100 μg of inhibitor μg |
|---|---|---|---|
| 28S | 7,460 | 670 | 9.0 |
| 18S | 3,250 | 450 | 13.8 |
| RNA III | 5,000 | 1,200 | 24.0 |
| 4S | 2,740 | 225 | 8.2 |

Radioactive RNA was obtained by injecting rats with 45 μCi of $^3H$ orotic acid per 100 g of body weight. RNA was prepared from liver nuclei by the procedure of Hiatti (1962) and Kruh (1967). 100 μg of each RNA fraction obtained on a sucrose gradient were incubated with 100 μg of reticulocyte inhibitor in the presence of 4 mM $MgCl_2$ at 4°C for 30 min. The inhibitor was precipitated at pH 5.0 by 1 N acetic acid and transferred to a millipore filter. The radioactivity of the filter was measured in a liquid scintillator. RNA III is the RNA fraction which sediments between 18S and 4S.

**Table 7.** Binding of liver $^3H$ RNA III with various protein fractions obtained from the reticulocyte pH 5 fraction on a DEAE cellulose column.

| Protein fraction | Radioactivity of 100 μg of RNA III c.p.m. | Radioactivity of the RNA-protein complex c.p.m. | Amount of RNA bound to 100 μg of protein μg |
|---|---|---|---|
| Eluate 0.02 M peak I | 5,000 | 1,200 | 24 |
| Eluate 0.02 M peak II | 5,000 | 294 | 5.9 |
| Eluate 0.03 M | 5,000 | 190 | 3.8 |

RNA III was prepared by centrifugation on a sucrose gradient. It sedimented between 18S and 4S. The protein fractions were obtained from reticulocyte pH 5 fraction first fractionated on CM cellulose, from a DEAE cellulose column by elution with 0.02 M tris buffer (pH 7.2) and with 0.03 M tris buffer (pH 7.2). The inhibitor is present in peak I of the 0.02 M eluate.

highest bound radioactivity was found with RNA III, which contains most of the messenger RNAs (Table 6). This is in agreement with the experiment described above concerning the action of the inhibitor on liver RNA in a sucrose gradient (Fig. 6). In the second experiment, RNA III was incubated with three protein fractions obtained from the DEAE-cellulose column, during the preparation of the reticulocyte inhibitor (Fig. 2). The peak I eluted by 0.02 M buffer which contains the inhibitor, was the most effective in binding to RNA (Table 7), 100 μg of reticulocyte inhibitor were able to bind to 24 μg of liver RNA III.

## BIOLOGICAL SIGNIFICANCE

It is not easy to understand at present the biological significance of these inhibitors. It could be postulated that these inhibitors are involved in the selection of proteins which are synthesized in differentiated cells. Gurdon (1968) has demonstrated that differentiated cells contain the complete information of the whole individual. One or several mechanisms should select the genes which will be expressed into proteins. Several observations have led to the hypothesis that a selection occurs at the transcription level which involves histone molecules (Huang and Bonner, 1962; Allfrey *et al.*, 1963). This selection is modulated by chromatin non histone proteins (Wang, 1967; Gilmour and Paul, 1969) and by protein kinase catalysed phosphorylation of histones (Kamiyama and Wang, 1971; Kamiyama and Dastugue, 1971). These mechanisms undoubtedly play an essential role in selecting the genes which will be transcribed, but they do not fit several observations which favor the idea that some messenger RNAs present in the cells are not translated. A very high molecular weight RNA has been found in the nucleus of avian erythroblasts but only a small fragment of it is present in polysomes (Attardi *et al.*, 1966; Scherrer *et al.*, 1966). A large part of the RNA present in the nucleus is destroyed locally, only a small part of it is transferred into cytoplasm in liver cell; the proportion of the transferred RNA varies with the condition of the cell (Shearer and McCarty, 1967; Drews *et al.*, 1968). We have found in liver, kidney and intestine nuclei an RNA fraction which sediments between 18S and 4S and which is likely to contain hemoglobin messenger RNA (Kruh *et al.*, 1964b, 1966).

If useless messenger RNAs are synthesized in differentiated cells, the inhibitors described in this paper could represent selection agents and play a regulatory role. These inhibitors can be related to other proteins involved in cell regulation: repressors of the genetic information (Gilbert and Muller-Hill, 1966; Ptashne, 1967; Riggs and Bourgeois, 1968) estradiol binding protein (Toft and Gorski, 1966). Common features of these inhibitors are that they constitute acidic proteins and are present in the cell in a very low concentration.

## ACKNOWLEDGEMENTS

These investigations were supported by "l'Institut National de la Santé et de la Recherche Medicale", "le Centre National de la Recherche Scientifique", "la Délégation Générale à la Recherche Scientifique et Technique", "le Commissariat à l'Energie Atomique" and "la Ligue Nationale Française contre le Cancer".

## REFERENCES

Allfrey, V. G., Littau, V. C. and Mirsky, E. A. (1963). *Proc. natn. Acad. Sci. U.S.A.* **49**, 414.
Attardi, G., Parnas, H., Wang, M. H. and Attardi, B. (1966). *J. molec. Biol.* **20**, 145.
Awdeh, Z. L., Williamson, A. R. and Askonas, B. A. (1968). *Nature, Lond.* **219**, 66.
Von Der Decken, A. and Campbell, P. N. (1962). *Biochem. J.* **84**, 499.
Drews, H., Brawerman, G. and Morris, H. P. (1968). *Eur. J. Biochem.* **3**, 284.
Gilbert, W. and Muller-Hill, B. (1966). *Proc. natn. Acad. Sci. U.S.A.* **56**, 1891.
Gilmour, R. S. and Paul, J. (1969). *J. molec. Biol.* **40**, 134.
Gurdon, J. B. (1968). *Scient. Am.* **219**, 24.
Hiatt, H. H. (1962). *J. molec. Biol.* **5**, 217.
Huang, R. C. and Bonner, J. (1962). *Proc. natn. Acad. Sci. U.S.A.* **48**, 1216.
Kamiyama, M. and Dastugue, B. (1971). *Biochem. biophys. Res. Commun.* **44**, 29.
Kamiyama, M. and Wang, T. Y. (1971). *Biochim. biophys. Acta* **228**, 563.
Kirby, K. S. (1956). *Biochem. J.* **64**, 405.

Kruh, J. (1967). *In* Methods in Enzymology (L. Grossman and K. Moldave eds.) Vol. 12A, p. 611, Academic Press, New York and London.
Kruh, J. (1968). *In* Methods in Enzymology (L. Grossman and K. Moldave, eds.) p. 728, Academic Press New York and London.
Kruh, J. and Levy, F. (1967). *Biochim. biophys. Acta* **145**, 460.
Kruh, J., Dreyfus, J. C. and Schapira, G. (1964a). *Biochim. biophys. Acta* **87**, 253.
Kruh, J., Dreyfus, J. C. and Schapira, G. (1964b). *Biochim. biophys. Acta* **91**, 494.
Kruh, J., Dreyfus, J. C. and Schapira, G. (1966). *Biochim. biophys. Acta* **114**, 371.
Labrie, F. (1969). *Nature, Lond.* **221**, 1217.
Levy, F., Tichonicky, L. and Kruh, J. (1970). *Biochim. biophys. Acta* **209**, 521.
Levy, F., Tichonicky, L. and Kruh, J. (1971). *Biochimie* **53**, 671.
Ptashne, M. (1967). *Proc. natn. Acad. Sci. U.S.A.* **57**, 306.
Riggs, A. D. and Bourgeois, S. (1968). *J. molec. Biol.* **34**, 361.
Schapiro, A. L., Vinuela, E. and Maizel, J. V. (1967). *Biochem. biophys. Res. Commun.* **28**, 815.
Scherrer, K., Marcaud, L., Zajdela, F., London, I. and Gros, F. (1966). *Proc. natn. Acad. Sci. U.S.A.* **56**, 1571.
Schweet, R. S., Lamfrom, H. and Allen, E. (1958). *Proc. natn. Acad. Sci. U.S.A.* **44**, 1029.
Shearer, R. W. and McCarty, J. B. (1967). *Biochemistry* **6**, 283.
Toft, D. and Gorski, J. (1966). *Proc. natn. Acad. Sci. U.S.A.* **55**, 1574.
Wang, T. Y. (1967). *J. Biol. Chem.* **42**, 1220.
Weber, K. and Osborn, M. (1969). *J. Biol. Chem.* **244**, 4406.
Wrighley, C. (1968). *Sci. Tools* **15**, 17.

FEBS Symposium, Volume 24, 1972, pp. 131-141

# Relationship between Nucleolar Size and the Synthesis and Processing of Pre-ribosomal RNA in the Liver of Rat

UNNE STENRAM

*Department of Pathology,*
*University of Uppsala, Uppsala* 1, *Sweden*

The main function of the nucleolus appears to be to synthesize ribosomal RNA, assemble it with protein and to process these ribonucleoprotein compounds into cytoplasmic ribosomal subunits (Perry, 1967; Busch and Smetana, 1970). After intraperitoneal administration of $^3$H-cytidine to rats, approximately 45 min elapse before labelled 28S (also denominated 29S) and 18S RNA of presumably ribosomal nature appear in the liver cell cytoplasm. This may be taken as a measure of the shortest time required for a normal liver cell nucleolus to make cytoplasmic ribosomal subunits. Thereby a method is offered by which to determine whether the nucleolus functions normally in the synthesis and processing of pre-ribosomal compounds. Little is known about further possible functions of nucleoli.

Protein-fed rats have larger liver cell nucleoli than starved rats, presumably due to an increased ribonucleoprotein synthesis (Lagerstedt, 1949). Animals fed a protein-free diet have still larger liver cell nucleoli (Stenram, 1956a), due to the absence of essential amino acids from the diet (Stenram, 1956b). This was an incitement to further investigations.

Interferometric and electronmicroscopic studies revealed that the liver nucleoli of rats fed with the protein-free diet are essentially normal as regards RNA and protein concentration, and ultrastructure (Stenram, 1966a). However, a slight modification in the ultrastructure was found, in agreement with other investigations, after prolonged protein deficiency. This was an increase in the granular component which is thought to contain 35-28S RNA, while 45S RNA is thought to be contained in the fibrillar area (Bernhard and Granboulan, 1968). Autoradiographic studies showed that the liver nucleolus and the whole cell of the protein-deprived rat have an increased RNA labelling after administration of $^3$H-cytidine (Stenram, 1962). This is, however, no measure of synthesis, as pool size and reutilization of products also affect the labelling. Later biochemical studies by Quirin-Stricker and Mandel (1968) revealed that the increased liver RNA labelling in protein deprivation is not

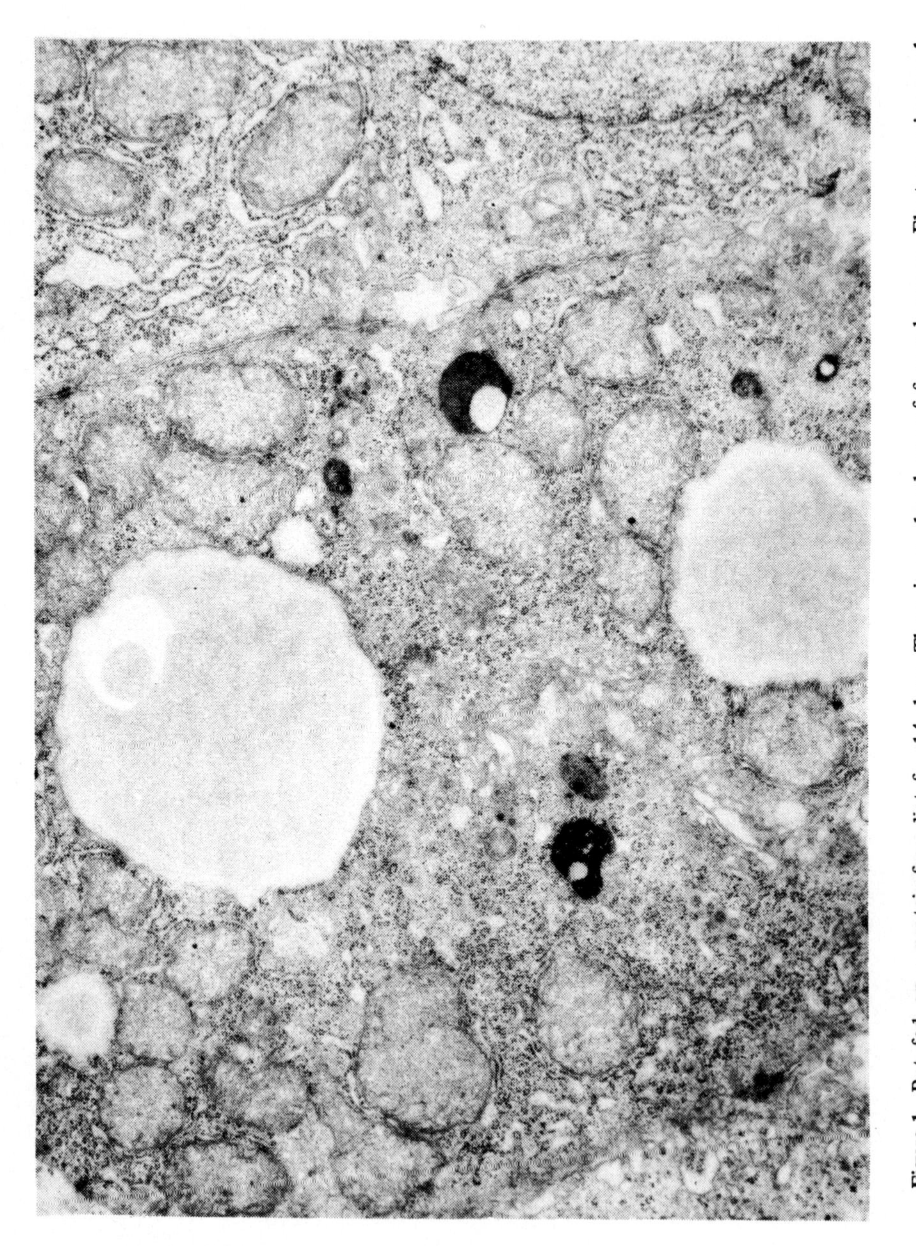

**Figure 1.** Rat fed on a protein-free diet for 14 days. There is an abundance of free polysomes. Electron micrograph of liver section. Uranyl acetate-lead citrate staining. × 15,000;

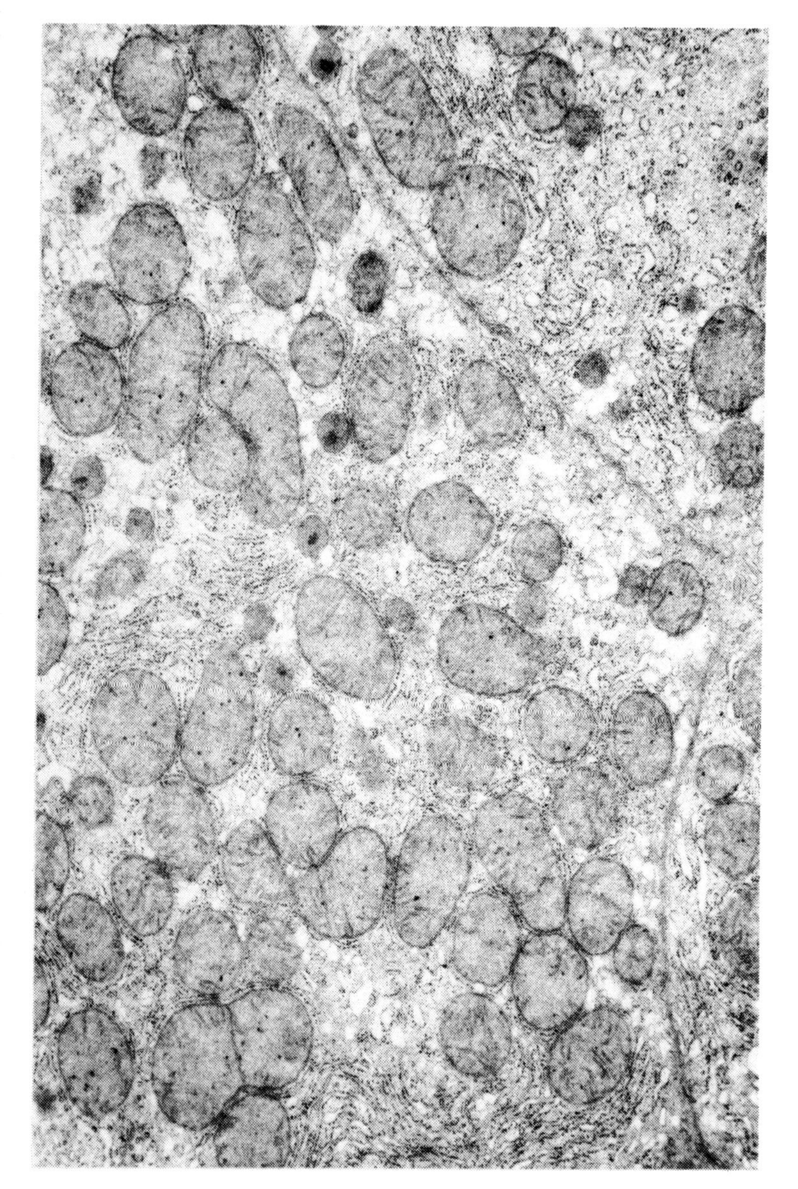

**Figure 2.** Rat pair-fed on a 25% casein diet for 14 days. The cytoplasm is fairly rich in rough-surfaced endoplasmic reticulum. Electron micrograph of liver section. Uranyl acetate-lead citrate staining. × 15,000.

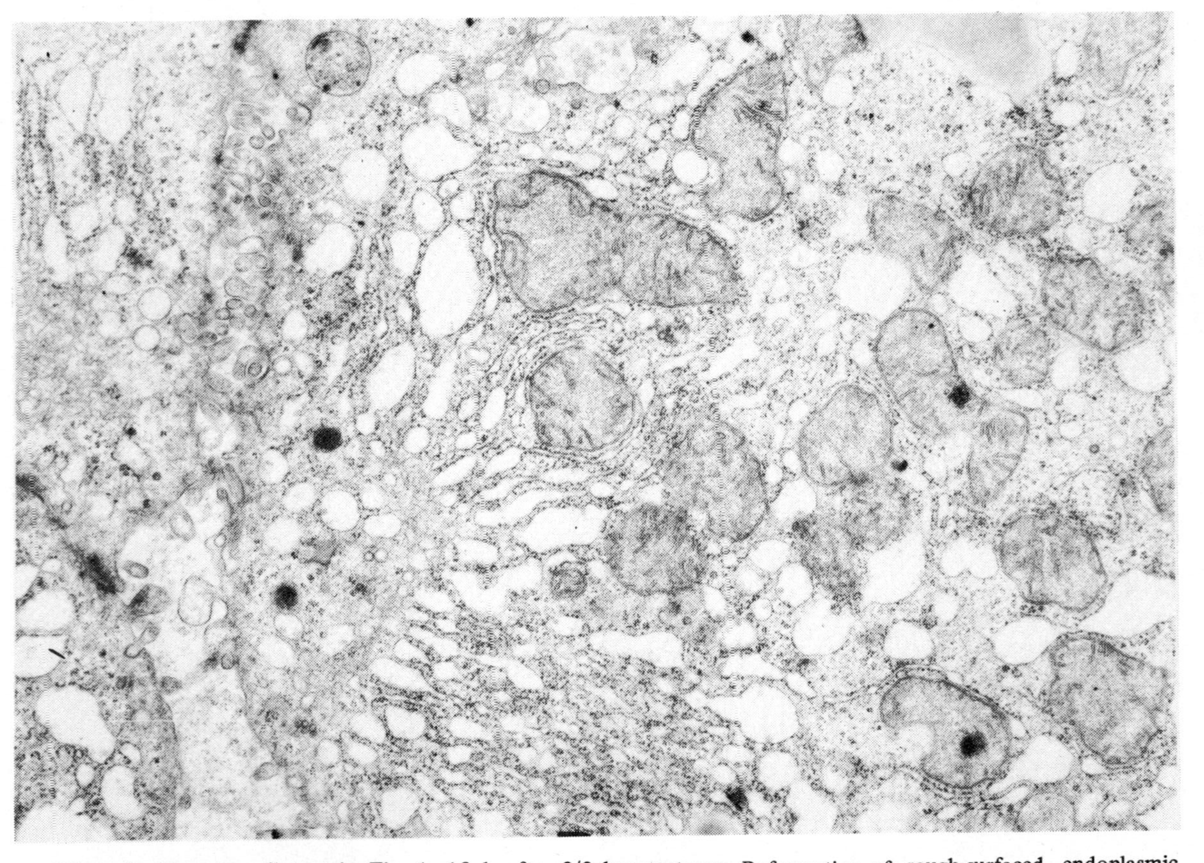

**Figure 3.** The same diet as in Fig. 1. 18 h after 2/3 hepatectomy. Reformation of rough-surfaced endoplasmic reticulum. Electron micrograph of liver section. Uranyl acetate-lead citrate staining. × 15,000.

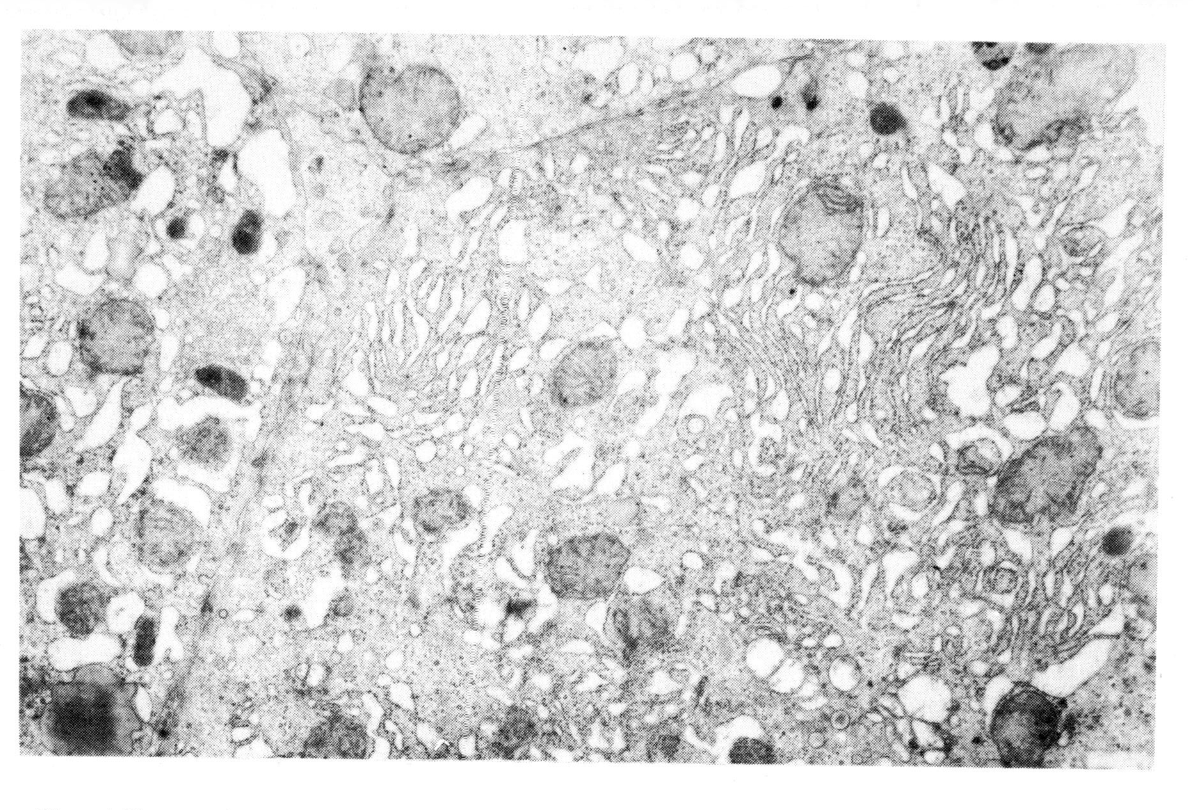

**Figure 4.** The same diet as in Fig. 2. 18 h after 2/3 hepatectomy. Reformation of rough-surfaced endoplasmic reticulum. For further information see Stenram *et al.*, 1970. Electron micrograph of liver section. Uranyl acetate-lead citrate staining. × 15,000.

due to altered pool size but represents increased synthesis. Shaw and Fillios (1968) found that liver RNA polymerase activity was increased in protein deprivation. There appeared to be no delay in the transport of labelled nucleolar pre-ribosomal RNA out into the cytoplasm under this condition (Stenram *et al.*, 1970). The liver nucleolar enlargement during protein deprivation may therefore be due to an increased synthesis in the nucleoli.

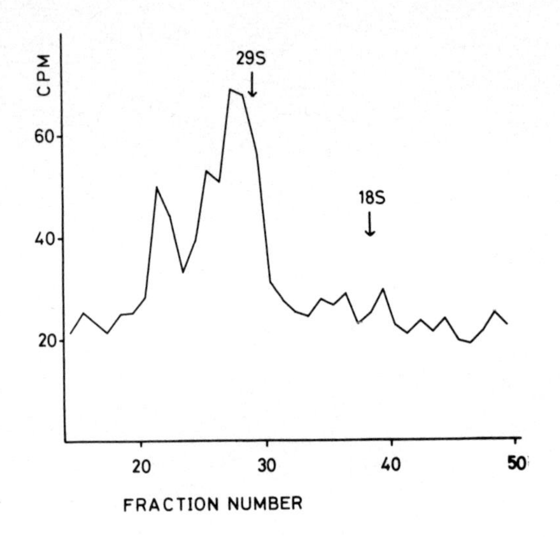


**Figure 5.** Nucleolar RNA from the liver of a rat given 5 μc 5-fluorouracil-6-H3 (Radiochemical Centre, Amersham, England, spec. activity 1.08 c/mmol) per g body weight intraperitoneally 3 h prior to death, followed after 10 min with 0.06 mg cold uracil, dissolved in ammoniac solution, pH 9.5, per g body weight. The exact localization of 29S and 18S was determined with the aid of $^{14}C$ labelled ribosomal RNA. The main peak is at 32S RNA. The other two peaks of heavy RNA may be 45S and 36S RNA. Labelling patterns of RNA following 2.0% polyacrylamide 0.5% agarose (w/w/v) gel electrophoresis. CPM = counts per min. For further technical details see Stenram *et al.*, 1972.

The reason for the increased synthesis is not known. In the liver cell cytoplasm, the rough-surfaced endoplasmic reticulum is decreased in amount in protein deficiency, but there is an abundance of free polysomes (Figs. 1-2). The half-life of the total cytoplasmic ribosomal RNA is decreased (Stenram and Nordgren, 1970, Enwonwu and Sreebny, 1971). It is not yet known to what extent this change affects free or membrane-bound ribosomes or both. The decrease in membrane-bound ribosomes with apparent preservation of free polysomes is also seen in rats given amino acid diets devoid of any one or of all of the essential amino acids (Stenram, 1971a).

Nucleolar enlargement is found in liver regeneration following partial hepatectomy (Stowell, 1948). The RNA polymerase activity is increased

(Tsukada and Lieberman, 1965). Nucleolar ultrastructure is essentially normal with a well developed granular component. The release of ribosomal RNA into the cytoplasm and reformation of ribosomes proceeds in a normal way whether the rat is given a protein-free or a casein-rich diet (Fig. 3.4) (Stenram *et al.*, 1970). The nucleolar enlargement is therefore probably due to an increased synthesis.

Rats fed on a thyroid-containing diet also develop enlarged liver cell nucleoli, considered to represent increased synthesis, although no detailed studies have been performed (Stenram, 1957).

The liver nucleoli enlarge after high doses of 5-fluorouracil (FU). The autoradiographic RNA labelling following $^3$H-cytidine administration is depressed in the whole cell. The nucleolar ultrastructure is altered with a special arrangement of the fibrillar, granular and electron-lucent components and of the individual granules (Stenram, 1966b). If lower doses are given or if the rats are killed soon after administration of the drug, changes are not seen in nucleolar ultrastructure. At low doses nucleoli are labelled following $^3$H-cytidine administration (Stenram and Willén, 1967). There is a delayed appearance of stable cytoplasmic ribosomal RNA labelling (Willén, 1970; Stenram *et al.*, 1972). Following partial hepatectomy, no reformation of cytoplasmic ribosomes is found in FU treated rats, though the nucleoli increase in size compared to sham operated animals (Stenram and Willén, 1970).

Administration of $^3$H-FU gives a low RNA labelling of nucleoli and cytoplasm. The labelling of heavy nucleolar RNA (Fig. 5) 3 h after administration of $^3$H-FU may suggest delayed intranucleolar processing compared to controls (Stenram *et al.*, 1972), but analysis of the time sequences before and after transcription has not been performed. Cytoplasmic 29S and 18S RNA were definitely labelled at 18 h, but not at 3 h (Fig. 6). Times in between have not been examined. The results suggest that, though delayed, FU-containing ribosomes enter the cytoplasm, and that the pre-ribosomal and 29S and 18S ribosomal RNA labelling obtained with $^3$H-cytidine after FU treatment (Stenram, 1966a, b; Stenram and Willén, 1967, 1970; Willén, 1970; Stenram *et al.*, 1972) represents FU-containing RNA. Base analysis has, however, not been performed, so we do not know, whether the labelled RNA contains FU or not.

The observations suggest that the nucleolar enlargement during treatment with FU are due to delayed release of stable ribosomal compounds into the cytoplasm.

Liver nucleolar enlargement is also provoked by treatment with thioacetamide (Rather, 1951). The nucleolar ultrastructure is essentially normal with predominance of the granular component (Salomon *et al.*, 1962). The half-life of 45S nucleolar RNA is increased (Steele, 1966). At a labelling time

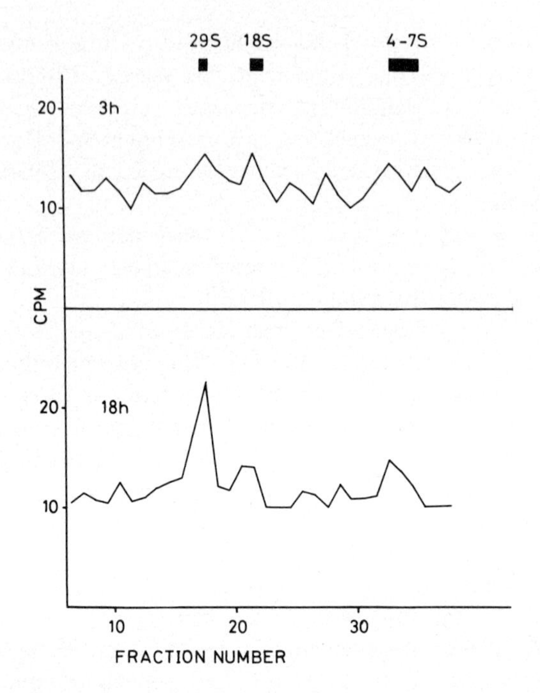


**Figure 6.** Cytoplasmic RNA from the liver of rats given 10 $\mu$c 5-fluorouracil-6-H3 per g body weight intraperitoneally 3 and 18 h before death. 29S and 18S RNA is labelled at 18 h. Due to the tenfold amount of RNA applied on the gels, the specific activity of the peaks should be reduced to a tenth when compared with the profiles of cytoplasmic RNA in Stenram *et al.*, 1972. Labelling patterns of RNA following 2.0% polyacrylamide 0.5% agarose (w/w/v) gel electrophoresis. CPM = counts per min. For further technical details see Stenram *et al.*, 1972.

of 45 min, we found good maturation into 32S RNA (Fig. 7). The entrance of ribosomal RNA into the cytoplasm is delayed (Stenram, 1971b). An obstacle may therefore exist at the transport to the cytoplasm. The delayed processing is probably one reason for the nucleolar enlargement, but the synthesis may also be increased (Steele, 1966).

Recently, we noted liver nucleolar enlargement in rats fed with a diet containing 4% rape-seed meal protein and 6% casein for 3 months (Abrahamsson *et al.*, 1971). The diet is sufficient in essential amino acids, though their availability may not be proved. The enlarged nucleoli were found in the central parts of the liver lobuli. They have a normal ultrastructure. The cytoplasm is rich in smooth-surfaced endoplasmic reticulum (Fig. 8). Rape-seed meal contains thioglucosides, and their break-down products, nitriles and goitrin, have been considered to affect the liver (VanEtten *et al.*, 1969). The

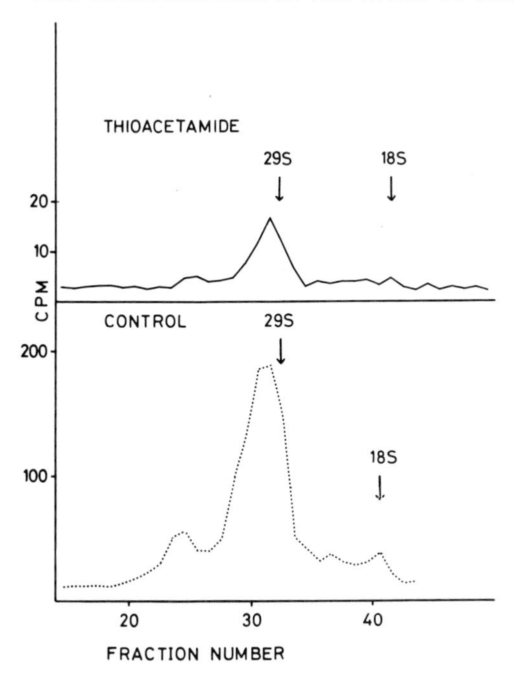


**Figure 7.** Liver nucleolar RNA of control and thioacetamide treated rats (50 mg per kg body weight intraperitoneally daily for 5 days: killed on the 6th day) 2.5 $\mu$c cytidine-$^3$H(G) (Radiochemical Centre, Amersham, England, spec. activity 2.8 c/mmol) was given intraperitoneally 45 min before death, and was followed after 10 min with 0.06 mg cold cytidine per g body weight. The specific activity is higher in the control rat. 32S RNA is labelled in both control and treated rat, as well as a minor peak at the 45S region. Labelling patterns of RNA following 2.0% polyacrylamide 0.5% agarose (w/w/v) gel electrophoresis. CPM = counts per min. For further technical details see Stenram *et al.*, 1972).

changes demonstrated were not found in rats fed with a diet containing 10% purified rape-seed meal protein.

The liver nucleoli decrease in size during treatment with actinomycin (Stenram, 1964), a condition in which nucleolar RNA transcription is largely inhibited (Perry, 1967). Already transcribed nucleolar RNA appears to mature in a normal way and eventually leaves the nucleolus, though stable cytoplasmic ribosomal RNA does not appear (Stenram *et al.*, 1972).

It appears at this time, that liver nucleolar enlargement could be caused by two different mechanisms–increased synthesis of pre-ribosomal RNA and delayed intranucleolar processing of the pre-ribosomal ribonucleoprotein compounds (the ribosomal proteins are probably synthesized in the cytoplasm and transported to the nucleoli). As these compounds seem to exist most of their life-time as 35-28S RNA containing nucleoprotein, the granular component, at

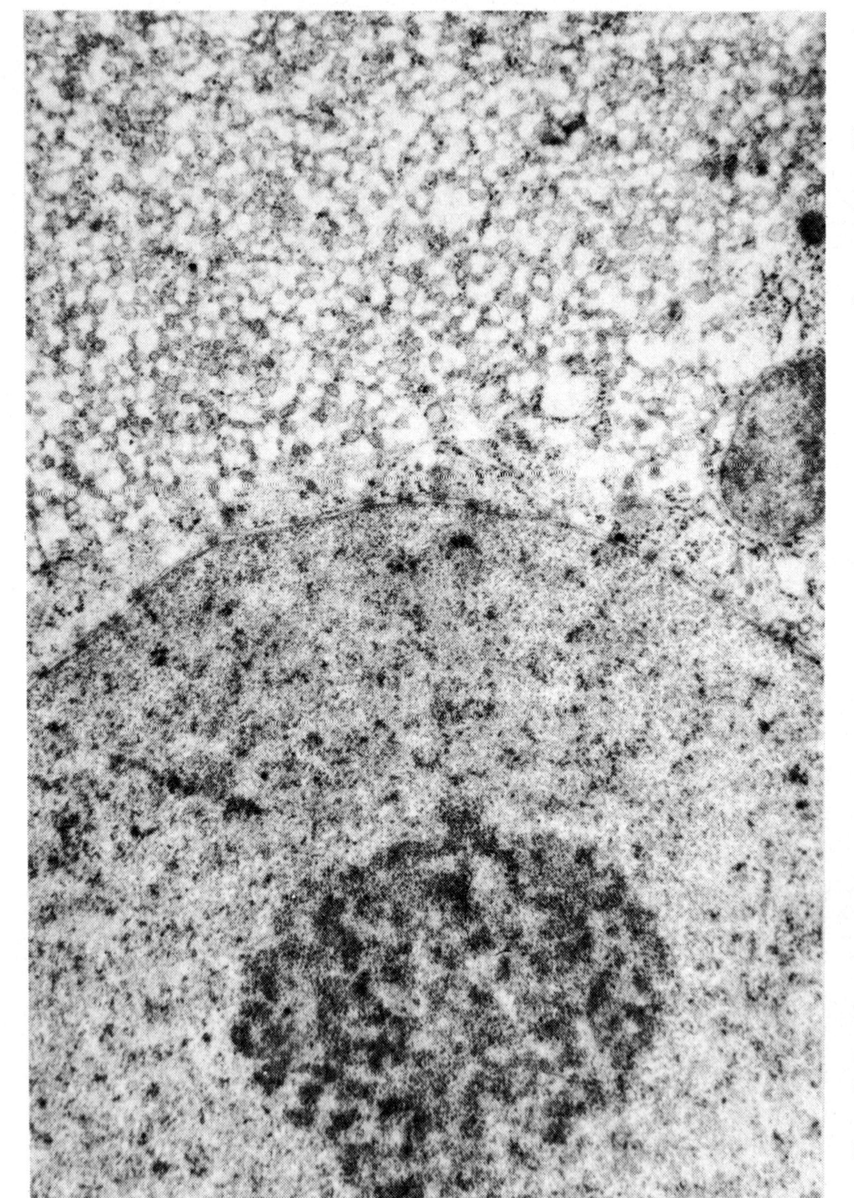

**Figure 8.** Electron micrograph of liver section. The rat was fed for 3 months with a diet containing 4% rape-seed meal protein and 6% casein. Note the large nucleolus and the abundance of smooth-surfaced endoplasmic reticulum. Uranyl acetate-lead citrate staining. × 16,750.

the ultrastructural level, may be considered to be especially prominent in increased synthesis, as in protein deficiency and after partial hepatectomy, as well as in delayed release of the final compounds from the nucleoli into the cytoplasm, as appears to be the situation after treatment with FU and thioacetamide. It remains to be determined to what extent other processes than those dealing with the pre-ribosomal compounds may be of importance for nucleolar size.

## REFERENCES

Abrahamsson, L., Hambraeus, L., Liedén, S.-Å., Nordgren, H. and Stenram, U. (1971). unpublished observations.

Bernhard, W. and Granboulan, N. (1968). *In* Ultrastructure in Biological Systems (A. Dalton and F. Haguenau, eds), Vol. 3, p. 81-149, Academic Press, New York and London.

Busch, H. and Smetana, K. (1970). The Nucleolus. Academic Press, New York and London.

Enwonwu, C. and Sreebny, L. (1971). *J. Nutr.* **101**, 501-514.

Lagerstedt, S. (1949). *Acta anat.* (Basel) suppl. IX.

Perry, R. P. (1967). *Progr. Nucl. Acid. Res. molec. Biol.* **6**, 219-257.

Quirin-Stricker, C. and Mandel, P. (1968). *Bull. Soc. Chim. biol.* **50**, 31-45.

Rather, L. J. (1951). *Bull. Johns Hopkins Hosp.* **88**, 38-58.

Salomon, J.-C., Salomon, M. and Bernhard, W. (1962). *Cancer Bull.* **49**, 139-158.

Shaw, C. and Fillios, L. C. (1968). *J. Nutr.* **96**, 327-336.

Steele, W. J. (1966). *In* Proceedings of the International Sympos. on the Cell Nucleus–Metabolism and Radiosensitivity, (H. M. Klouwen, ed.) p. 203-220, Taylor and Francis, London.

Stenram, U. (1956a). *Acta anat. (Basel)* **26**, 350-359.

Stenram, U. (1956b). *Acta path. microbiol. Scand.* **38**, 364-374.

Stenram, U. (1957). *Acta path. microbiol. Scand.* **40**, 407-412.

Stenram, U. (1962). *Z. Zellforsch. mikrosk. Anat.* **58**, 107-124.

Stenram, U. (1964). *Expl Cell Res.* **36**, 242-255.

Stenram, U. (1966a). *Natl. Cancer Inst. Monogr.* **23**, 379-390.

Stenram, U. (1966b). *Z. Zellforsch. mikrosk. Anat.* **71**, 207-216.

Stenram, U. (1971a). *Virchows Arch. Abt. B Zellpath.* **8**, 124-132.

Stenram, U. (1971b). *Hoppe-Seyler's Z. physiol. Chem.* **352**, 674-682.

Stenram, U., Bengtsson, A. and Willén, R. (1972). *Cytobios* **5**, 125-143.

Stenram, U. and Nordgren, H. (1970). *Cytobios* **2**, 265-275.

Stenram, U., Nordgren, H. and Willén, R. (1970). *Virchows Arch. Abt. B Zellpath.* **6**, 12-23.

Stenram, U. and Willén, R. (1967). *Z. Zellforsch. mikrosk. Anat.* **82**, 270-281.

Stenram, U. and Willén, R. (1970). *Chem.-Biol., Interactions* **2**, 79-88.

Stowell, R. (1948). *Archs Path.* **46**, 164-178.

Tsukada, K. and Lieberman, I. (1965). *J. biol. Chem.* **240**, 1731-1736.

VanEtten. C. H., Gagne, W. E., Robbins, D. J., Booth, A. N., Daxenbichler, M. E. and Wolff, I. A. (1969). *Cereal Chem.* **46**, 145-155.

Willén, R. (1970). *Hoppe-Seyler's Z. physiol. Chem.* **351**, 1141-1150.

FEBS Symposium, Volume 24, 1972, pp. 143-145

# Selective Effects of α-amanitin on RNA Labelling in Explanted Salivary Glands of *Chironomus thummi* larvae

EDGAR SERFLING, ULRICH WOBUS, and REINHARD PANITZ

*Zentralinstitut für Genetik und Kulturpflanzenforschung der Deutschen Akademie der Wissenschaften, Gatersleben, DDR*

I would like to make a few comments on RNA synthesis in polytene chromosomes of the larval salivary gland of the midge *Chironomus thummi* under the influence of α-amanitin.

It is well established that α-amanitin *in vitro* inhibits the activity of only one of at least two RNA polymerase activities of eukaryotic cells (Kedinger *et al.*, 1970; Lindell *et al.*, 1970) by binding to the enzyme molecule itself (Meihlac *et al.*, 1970), thus preventing the synthesis of high molecular weight chromosomal RNA. Ribosomal RNA synthesis remains unimpaired. After administration of the compound to whole animals (Niessing *et al.*, 1970; Jacob *et al.*, 1970), intact cells (Seitz and Seitz, 1971) or oocytes (Tocchini-Valentini and Crippa, 1970), however, contradictory results were obtained. In explanted salivary glands of *Chironomus thummi*, α-amanitin specifically inhibits chromosomal (or puff) RNA synthesis, as shown by autoradiographic experiments (Wobus *et al.*, 1971) (Fig. 1, insets). A more detailed analysis of the RNA species synthesized by chromosomes and nucleolus, and isolated by micromanipulation revealed results demonstrated in Fig. 1. As may be seen, the most obvious effect of the toxin is the complete absence of high molecular weight chromosomal RNA, a result obtained simultaneously by the Edström group (personal communication). Some bands on the chromosomes, the labelling of which remains unimpaired by amanitin (cf. Fig. 1, lower inset), seem to represent sites of tRNA synthesis (Edström, personal communication). The synthesis of RNA in the 4-5S range is not affected. All the RNA species synthesized in the nucleoli of the control are also formed in the nucleoli of amanitin-treated glands but in amounts differing from those in untreated glands. The same radioactivity profile obtained after 4 h synthesis in the presence of amanitin is available from the control after only one hour of incubation. Therefore, it could be assumed that the toxin in all probability

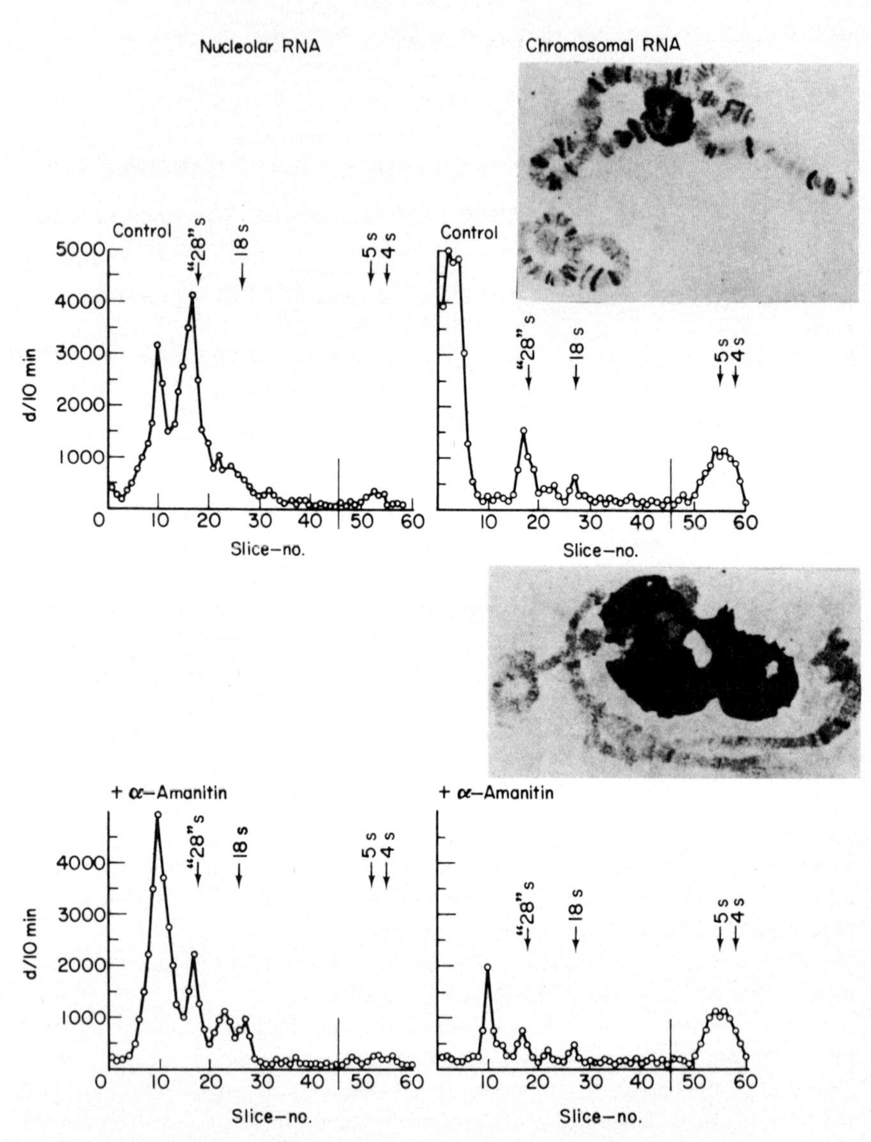


**Figure 1.** Separation of nucleolar and chromosomal RNA after labelling for 4 h in the absence (control) and in the presence of α-amanitin. *Chironomus thummi* prepupal salivary glands were preincubated at room temperature for 30 min in 20 μl Cannon's modified insect medium (Ringborg *et al.*, 1970) containing 1 μg/ml α-amanitin, followed by a further 4 h incubation in the same medium supplemented with 100 μCi $^{3}$H-cytidine (spec. act. 18.8 Ci/mmole) and 100 μCi $^{3}$H-uridine (spec. act. 24.5 Ci/mmole). Control glands were incubated for 4 h under the same conditions in amanitin-free medium containing $^{3}$H-cytidine and $^{3}$H-uridine as specified above. The glands were fixed, nucleoli (30-50) and chromosomes (70-100) isolated by micromanipulation and the RNA

does not impair the formation of the 38S rRNA precursor but in some way inhibits the processing of this RNA species.

Another remarkable effect is the appearance of RNA on the chromosomes in the position of the 38S rRNA precursor in addition to 30S, 23S and 18S RNA, the only species found on the chromosomes in control glands. This may be interpreted as a further indication for disturbances by α-amanitin in the processing and, possibly, transport of rRNA precursors.

Our experimental results confirm the specificity of action of α-amanitin in an explanted insect tissue and support the suggestion that 4-5S RNA is transcribed by a third polymerase activity (Blatti *et al.*, 1971). Additionally, they demonstrate the impairment by α-amanitin of the nucleolus function at the post-transcriptional level.

## REFERENCES


Bishop, D. H. L., Claybrook, J. R. and Spiegelman, S. (1967). *J. molec. Biol.* **26**, 373.

Blatti, S. P., Ingles, C. J., Lindell, T. J., Morris, P. W., Weaver, R. F. Weinberg, F. and Rutter, W. J. (1971). *Cold Spring Harb. Symp. quant. Biol.* **35**, 649.

Edström, J.-E. and Daneholt, B. (1967). *J. molec. Biol.* **28**, 331.

Jacob, S. T., Muecke, W., Sajdel, E. M. and Munro, H. N. (1970). *Biochem. biophys. Res. Commun.* **40**, 334.

Kedinger, C., Gniazdowski, M., Mandel, J. C. Jr., Gissinger, F. and Chambon, P. (1970). *Biochem. biophys. Res. Commun.* **38**, 165.

Lindell, T. J., Weinberg, F., Morris, P. W., Roeder, R. G. and Rutter, W. J. (1970). *Science, N.Y.* **170**, 447.

Meihlac, M., Kedinger, C., Chambon, P., Faulstich, H., Govindan, M. V. and Wieland, T. (1970). *FEBS Letters* **9**, 258.

Niessing, J., Schnieders, B., Kunz, W., Seifart, K. H. and Sekeris, C. E. (1970). *Z. Naturf.* **25b**, 1119.

Ringborg, U., Daneholt, B., Edström, J.-E., Egyhazi, E. and Rydlander, L. (1970). *J. molec. Biol.* **51**, 679.

Seitz, U. and Seitz, U. (1971). *Planta* **97**, 224.

Tocchini-Valentini, G. P. and Crippa, M. (1970). *Nature, Lond.* **228**, 993.

Wobus, U., Panitz, R. and Serfling, E. (1971). *Experientia* **27**, 1202.


---

extracted by pronase-SDS-treatment (Edström and Daneholt, 1967). Cold RNA of *Ch.thummi* larvae was used as carrier and marker. Electrophoresis was run according to Bishop *et al.*, (1967) in composite gels containing 2.4% (4.5 cm) and 7.2% (3 cm) acrylamide, respectively. In the figure, the border between the two parts of the gel is marked by a vertical line between slice-numbers 40 and 50. Under our experimental conditions the correct S-values of 28S rRNA and its precursors are not 38S, 30S and 28S as indicated in the text but approximately 36S, 27S and 26S, respectively (E. Serfling in preparation). Insets: autoradiographs of orcein-acetic acid stained chromosome glands treated as described above.

FEBS Symposium, Volume 24, 1972, pp. 147-149

# RNA Synthesis in Salivary Gland Chromosomes of *Chironomus thummi*

E. SERFLING, U. WOBUS and R. PANITZ

*Zentralinstitut für Genetik und Kulturpflanzenforschung der Deutschen Akademie der Wissenschaften, Gatersleben, DDR*

I want to add some remarks about our work on RNA synthesis in giant chromosomes of *Chironomus thummi* performed by E. Serfling, U. Wobus and myself. For the experiments we used the salivary gland chromosomes which were allowed to incorporate tritiated uridine and cytidine by incubating the glands in a synthetic medium. The distribution of RNA isolated from microdissected nucleoli and chromosomes was studied by electrophoresis on polyacrylamide gels.

After 1 h incubation the labelled RNA of isolated nucleoli contains, besides a main fraction of about 38S, three smaller peaks at the 30S, 23S and 18S position (Fig. 1). With longer incubation times (4 and 16 h) two large components become the dominating fractions in a proportion of nearly 1:1, while the 18S is now absent. The 30S and 23S fractions are likely to be precursors of the 28S and 18S rRNA. Our results are in good agreement with those of Ringborg and co-workers (Ringborg *et al.*, 1970a) with the exception that these authors failed to detect 18S RNA within the nucleolus and thus concluded that the maturation to the finished rRNA molecules is an extra-nucleolar process. It is unclear to what extent this difference is due to biological or technical reasons. The absence of 18S RNA after long incubation times in our experiments is difficult to explain. Several authors (Soeiro *et al.*, 1968; Ringborg and Rydlander, 1971) found 18S rRNA to be more sensitive to exogenous damage than 28S rRNA. Therefore, its absence may be caused by the unfavourable experimental conditions of long incubation times.

At all incubation times the radioactivity profile of chromosomes I-III revealed a high molecular weight fraction of more than 60S. The activity of this fraction increased between 1 and 4 h while its profile remained unchanged, suggesting no conversion on the chromosomes. In addition, after 4 h 30S and 18S were obtained on the chromosomes. The delayed labelling of these fractions suggests their origin from preribosomal RNA. This conclusion

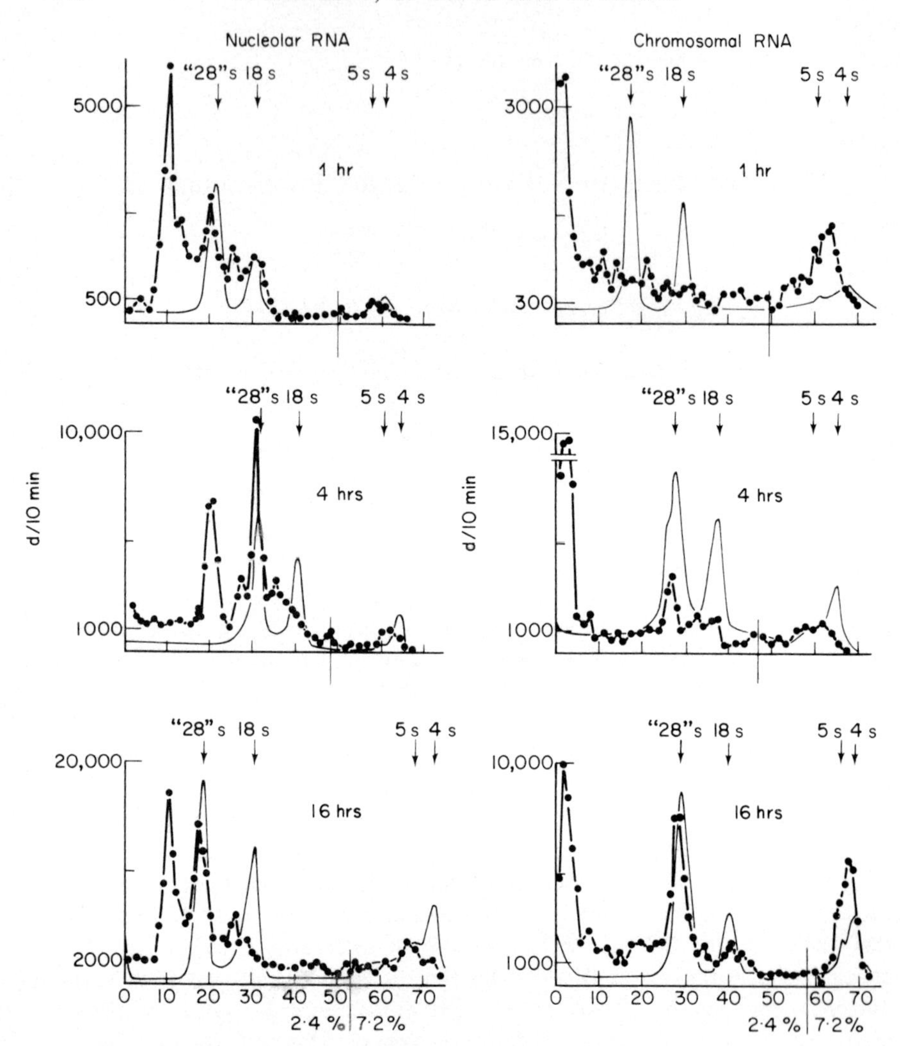


**Figure 1.** Separation of *in vitro* labelled nucleolar and chromosomal RNA after different incubation times. *Chironomus thummi* prepupal salivary glands were incubated with 100 $\mu$C $^{3}$H-cytidine (spec. act. 18.8 Ci/mmol) and 100 $\mu$C $^{3}$H-uridine (spec. act. 24.5 Ci/mmol) in 20 $\mu$l Cannon's modified insect medium (Ringborg *et al.*, 1970b) at room temperature. The glands were fixed, nucleoli (30-50) and chromosomes I-III (70-100) isolated by micromanipulation and the RNA extracted by pronase-SDS-treatment according to Edström and Daneholt (1967). Cold RNA of *Ch. thummi* larvae was used as carrier and marker. Electrophoresis was performed according to Bishop *et al.* (1967) in composite gels containing 2.4% (4.5 cm) and 7.2% (3 cm) acrylamide, respectively. In the figure the border between the two parts of the gel is marked. Under our experimental conditions the correct S-values of 28S rRNA and its precursors are not 38S, 30S and 28S as indicated in the text but approximately 36S, 27S and 26S, respectively (Serfling, in preparation).

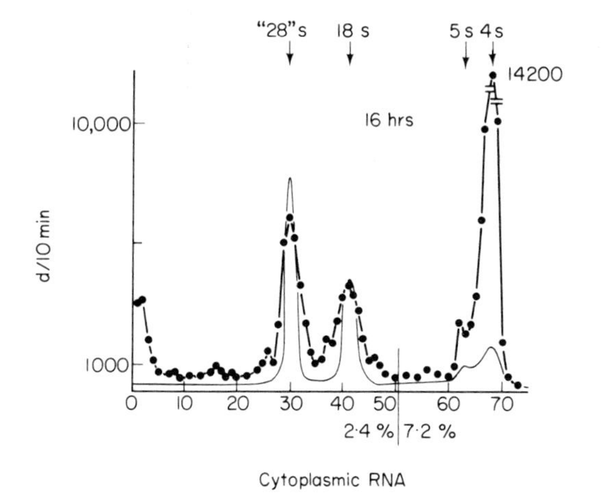


**Figure 2.** Separation of *in vitro* labelled cytoplasmic RNA. For other details see Fig. 1.

is also reinforced by the fact that both fractions are obtained on the chromosomes after inhibiting by α-amanitin the synthesis of high molecular weight chromosomal RNA only (Serfling, *et al.* (1973). Furthermore, the chromosomes contain low molecular weight RNA in the range of 4-5S. With increasing incubation time a shift occurs in this part of the profile suggesting a conversion of precursor molecules.

The radioactive profile of cytoplasmic RNA (Fig. 2) gives some information about the kinetics of RNA synthesis. A comparison of the relative proportion of label between the ribosomal RNAs and 4S RNA revealed the highest amount in the latter. This indicates that 4S RNA migrates faster to the cytoplasm than ribosomal RNAs. The diagram also shows that under our experimental conditions a normal processing of rRNAs takes place as demonstrated by the identity of the radioactivity and optical density profiles.

## REFERENCES

Bishop, D. H. L., Claybrook, J. R. and Spiegelman, S. (1967). *J. molec. Biol.* **26**, 373.

Edström, J.-E. and Daneholt, B. (1967). *J. molec. Biol.* **28**, 331

Ringborg, U. and Rydlander, L. (1971). *J. Cell Biol.* (in press).

Ringborg, U., Daneholt, B., Edström, J.-E., Egyházy, E. and Lambert, B. (1970a). *J. molec. Biol.* **51**, 327.

Ringborg, U., Daneholt, B., Edström, J.-E., Egyházy, E. and Rydlander, L. (1970b). *J. molec. Biol.* **51**, 679.

Serfling, E., Wobus, U. and Panitz, R. (1973). This volume, p. 143.

Soeiro, R., Vaughan, M. H. and Darnell, J. E. (1968). *J. Cell Biol.* **36**, 91.

# 15

FEBS Symposium, Volume 24, 1972, p. 151

# Synthesis of Nucleic Acids in Sea Urchin Development

GIOVANNI GIUDICE

*Institute of Comparative Anatomy*
*The University of Palermo, Italy.*

## STUDY OF RIBOSOMAL RNA SYNTHESIS DURING SEA URCHIN DEVELOPMENT

Ribosomal RNA in sea urchin gastrulae is synthesized in the form of a 33S precursor (of $2.58 \times 10^6$ daltons) that is quickly cleaved into the mature 18S subunit and into a 28S fragment. The latter slowly matures into the 26S ribosomal RNA subunit.

Methods have been developed to isolate clean sea urchin oocytes in bulk amounts. These observations prove that maturation of the ribosomal RNA precursor occurs at a much slower rate than during embryogenesis.

The possibility is considered that the rate of precursor maturation represents one step at which the rate of production of mature ribosomal RNA is regulated.

## REGULATION OF DNA SYNTHESIS IN SEA URCHIN EMBRYONIC CELLS

Sea urchin embryos can be dissociated into single cells.

These latter, under appropriate conditions, are able to reaggregate and to differentiate again into larvae-like aggregates.

When cells dissociated from blastulae are prevented from reaggregating by overdilution, DNA synthesis is almost completely and immediately halted, whereas if cells are permitted to reaggregate, DNA synthesis continues to proceed.

The possibility has been considered that in the absence of intercellular contacts a "signal" originating in the membrane is responsible for the inhibition of DNA synthesis. This suggestion is supported by the finding that a mild treatment with trypsin of the non-aggregating cells partially restores DNA synthesis.

FEBS Symposium, Volume 24, 1972, pp. 153-158

# The Hormonal Receptors of Chromatin and Their Probable Role in Ontogenesis

R. I. SALGANIK, T. M. MOROZOVA and I. A. LAVRINENKO

*Institute of Cytology and Genetics,*
*Siberian Branch of the U.S.S.R. Academy of Sciences,*
*Novosibirsk, U.S.S.R.*

It is becoming increasingly clear that most hormones, in consequence of their interaction with specific receptors, accumulate in corresponding target cells (Greenspan and Hargadine, 1965; Fanestil and Edelman, 1966; Noteboom and Gorski, 1965). Receptors for different hormones were found in plasma membranes (Rodbell *et al.*, 1970; Cautrecasas, 1971), in cytoplasm and in nuclei of target cells (Noteboom and Gorski, 1965; Toft *et al.*, 1967; Yensen *et al.*, 1968; Morozova and Salganik, 1969; Swaneck *et al.*, 1970). Recently we have shown that hormones acting as inductors of transcription interact specifically with the chromatin of target tissues. This question was examined in our laboratory by using the technique of equilibrium dialysis (Lavrinenko *et al.*, 1971). It has been found that the binding of labelled oestradiol by rat uterus chromatin is much higher than by kidney chromatin, (Fig. 1) whereas kidney chromatin binds 1,2 $^3$H aldosterone more intensively than uterus and liver chromatin (Fig. 2) 1,2 $^3$H-cortisone was bound equally well by liver and kidney chromatin presumably because both organs serve as targets for corticosteroids. Non-labelled aldosterone competes with labelled aldosterone for kidney chromatin, but does not prevent 6,7$^3$H-oestradiol binding while diethylstilbestrol competes with 6,7$^3$H-oestradiol for uterus chromatin (Table 1). These data may be considered as strong evidence for the high specificity of hormone-chromatin interaction, suggesting the existence of receptors for corresponding hormones in target cell chromatin. It appears that such hormone receptors are connected with the fraction of non-histone proteins of chromatin. When 1,2$^3$H-cortisol was administered *in vivo* it was found only in non-histone proteins of rat liver chromatin, and not in histones or DNA (Morozova and Salganik, 1969). The extraction of histones did not alter the ability of kidney and uterus chromatin to bind selectively aldosterone or oestradiol, respectively.

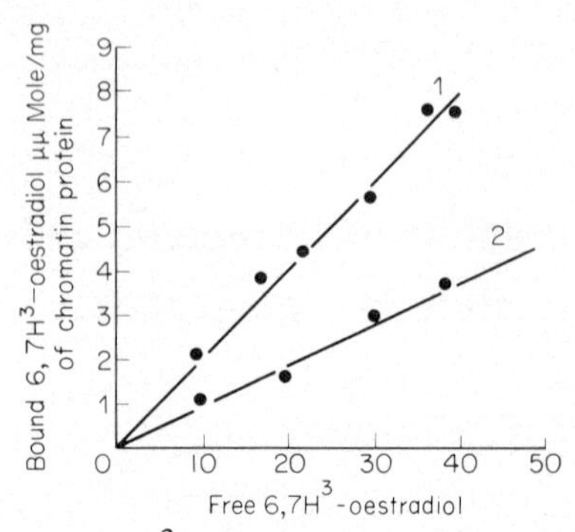


**Figure 1.** The binding of 6,7 $^{3}$H-oestradiol by uterus and kidney chromatin. The binding of hormones to chromatin was studied by the method of equilibrium dialysis. Chromatin solution in distilled water (0.5-1.0 mg of chromatin protein/ml) was placed in a cell and an equal volume of distilled water was poured out in the adjacent one. To both cells, separated by cellophane membrane, equal amounts of labelled hormone (10-50 $\mu\mu$mole/ml) were added. Dialysis was carried on for 72 h at 4°C. The distribution of labelled hormone between cells was measured. Aliquots of 0.2 ml were counted in a scintillation counter Mark I (Nuclear Chicago). 1. Uterus chromatin; 2. Kidney chromatin.

**Table 1.** Competition of unlabelled steroid hormones with 1,2 $^{3}$H-aldosterone and 6,7 $^{3}$H-oestradiol for target-tissue chromatin

| Experimental conditions | Binding of labelled hormone by chromatine ($\mu\mu$Mole/mg of protein) | | Binding inhibition% |
|---|---|---|---|
| | Control | Experiment | |
| Uterus chromatin + 6,7 $^{3}$H-oestradiol + diethylstylbestrol | 8.0 ± 0.6 | 3.0 ± 0.2 | 67 |
| Uterus chromatin + 6,7 $^{3}$H-oestradiol + aldosterone | 8.1 ± 0.5 | 8.2 ± 0.6 | 0 |
| Kidney chromatin + 1,2 $^{3}$H-aldosterone + aldosterone | 4.2 ± 0.3 | 1.34 ± 0.1 | 70 |
| Kidney chromatin + 6,7 $^{3}$H-oestradiol + aldosterone | 4.4 ± 0.3 | 4.0 ± 0.1 | 8 |
| Kidney chromatin + 1,2 $^{3}$H-aldosterone + hydrocortisone | 4.0 ± 0.2 | 2.9 ± 0.1 | 28 |

The binding of hormones with chromatin was studied by the method of equilibrium dialysis. To the solution of chromatin (1 mg of chromatin protein/ml) 80-100 $\mu\mu$mole labelled and 10 times as much unlabelled hormones were added. The mixture was dialysed against distillated water for 48-72 h at 2-5°C. In controls unlabelled hormone was omitted.

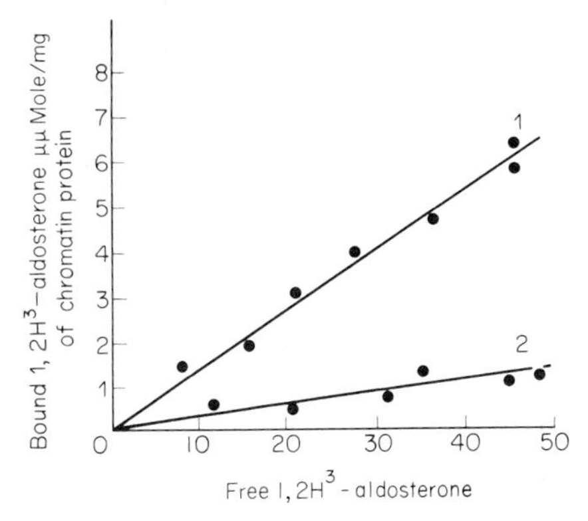


**Figure 2.** The binding of 1,2$^3$H-aldosterone by uterus and kidney chromatin. The experiments were performed as described in Fig. 1. Free 1,2$^3$H-aldosterone $\mu\mu$mole/ml 1. Kidney chromatin; 2. Uterus chromatin.

It is well known that certain hormones play an important role in the ontogenesis of animals acting as regulators of transcription for definite genes which provide differentiation and morphogenesis (Karlson, 1963; Tata, 1970). One would assume that the formation of hormone receptors in the chromatin of corresponding cells endows them with sensitivity to the appropriate hormones and may serve as an essential mechanism of development. The replacement of receptors in these cells may, in turn, contribute to changes in the development programme. Our investigation was aimed to test this assumption.

**Table 2.** Binding of 1,2 $^3$H-hydrocortisone by rat liver chromatin incubated with hormone

| Exp. No. | Concentration of 1,2 $^3$H-hydro-cortisone ($\mu\mu$Mole/ml) | Amount of bound 1,2 $^3$H-hydrocortisone (m$\mu\mu$Mole/mg chromatin DNA) | |
|---|---|---|---|
| | | Rat embryos | Adult rats |
| 1 | 8.3 | 34 | 133 |
| 2 | 10.0 | 24 | 67 |
| 3 | 26.2 | 184 | 900 |
| 3a | 26.2 | 163 | 557 |

Free and chromatin-bound 1,2$^3$H-hydrocortisone were separated by gel-filtration on Sephadex G-25. In Exp. 3a the chromatin from Exp. 3 was repeatedly subjected to gel-filtration. Rat liver chromatin was prepared according to Marushige and Bonner (1966). To the solution of chromatin in de-ionized water (1 mg of chromatin DNA/ml) 1,2$^3$H-hydrocortisone was added (8.27 $\mu\mu$Mole/ml) and incubated for 5 h at 4°C.

**Table 3.** Binding of 1,2 $^3$H-hydrocortisone by rat liver chromatin from liver homogenates incubated with hormone

| Exp. No. | Concentration of 1,2 $^3$H-hydro-cortisone ($\mu\mu$Mole/ml) | Amount of bound 1,2 $^3$H-hydrocortisone (m$\mu\mu$Mole/mg chromatin DNA) Rat embryos | 4 day old rats | Adult rats |
|---|---|---|---|---|
| 1 | 27 | 5.3 | – | 70.7 |
| 2 | 27 | 7.0 | – | 73.0 |
| 3 | 10 | 2.3 | – | 35.0 |
| 4 | 27 | – | 25.0 | 55.0 |
| 5 | 10 | – | 11.0 | 35.0 |

Livers were homogenized in 4 volumes of a medium containing 0.25 M sucrose, 25 mMKCl, 10 mMMgCl$_2$, 50 mM Tris (pH 7,6) and filtrated through nylon. To the filtrate 1,2$^3$H-hydrocortisone (10-27 $\mu\mu$Mole/ml) was added and the mixture was incubated for 15 min at 37°C. Chromatin was isolated from homogenate after this incubation according to Marushige and Bonner (1966).

Glucocorticoids are able to induce glycogenic enzymes in the liver of postnatal rats and practically do not act as inductors of these enzymes in rat embryo (Sereni, *et al.*, 1959; Yeung *et al.*, 1967). It appeared reasonable to suppose that the unresponsiveness of embryonic liver cells to glucocorticoids depends on the absence of chromatin receptors, which, probably, develop only in the postnatal period ensuring thereby cell reactivity to these hormones. We have investigated the existence of such receptors in the liver chromatin of pre- and postnatal rats.

It appears that liver chromatin from 19-20 days old rat embryos binds 1,2$^3$H-hydrocortisone 3-5 times less, compared with chromatin isolated from

**Table 4.** Binding of 1,2 $^3$H-hydrocortisone by rat liver cytoplasm

| Exp. No. | Amount of bound 1,2 $^3$H-hydrocortisone (m$\mu\mu$moles/g tissue) Embryos | Adult rats |
|---|---|---|
| 1 | 2270 | 3500 |
| 2 | 2150 | 3700 |
| 3 | 2720 | 3900 |

Rat livers were homogenzied in 4 volumes of a medium containing 0.25 M sucrose, 25 mMKCl, 10 mMMgCl$_2$, 50 mM Tris (pH 7.6) and centrifuged at 800 x g for 10 min. The resulting supernatant was recentrifuged at 145.000 x g for 90 min. To the supernatant 1,2$^3$H-hydrocortisone was added (27 $\mu\mu$Mole/ml) and the mixture was incubated for 5 h at 4°C. The free and chromatin-bound 1,2$^3$H-hydrocortisone were separated by gel-filtration on Sephadex G-25.

**Table 5.** Competition of unlabelled steroids with 1,2 $^3$H-hydrocortisone for rat liver chromatin and cytoplasmic receptors

| Exp. No. | | Binding of labelled hormone (mμμmole/mg chromatin DNA or cytoplasmic proteins) Control | Experiment | Binding inhibition % |
|---|---|---|---|---|
| 1. | Chromatin + 1,2 $^3$H-hydrocortisone + corticosterone | 71 | 25 | 65 |
| 2. | Chromatin + 1,2 $^3$H-hydrocortisone + aldosterone | 71 | 52 | 27 |
| 3. | Cytoplasm + 1,2 $^3$H-hydrocortisone + hydrocortisone | 125 | 36 | 71 |
| 4. | Cytoplasm + 1,2 $^3$H-hydrocortisone + testosterone | 125 | 61 | 51 |

Rat liver homogenates in Exp. 1 and 2 were prepared as described in Table 3. To the homogenates labelled hydrocortisone ($27.10^{-9}$ M) and unlabelled steroids ($10^{-6}$ M) were added and the mixtures were incubated for 15 min at 37°C. Chromatin from homogenates was isolated according to Marushige and Bonner (1966). Rat liver cytoplasm in Exp. 3 and 4 was prepared as described in Table 4. To the resulting supernatants labelled hydrocortisone ($27.10^{-9}$ M) and unlabelled steroids were added and the mixtures were incubated for 5 h at 4°C. The free and protein-bound hormones were separated by gel-filtration on Sephadex G-25.

adult rat liver (Table 2). This difference increases when tissue homogenates are incubated with labelled hydrocortisone before the isolation of chromatin (Table 3). In this case 1,2$^3$H-hydrocortisone content is 10-14 times lower in liver chromatin of rat embryos as compared with the liver chromatin from adult rats. This difference depends probably on the existence of intermediary cytoplasmic receptors, which transfer hydrocortisone to the receptors of chromatin thus enhancing the specificity of binding.

As a matter of fact, cytoplasmic supernatant of rat liver homogenate binds labelled hydrocortisone and the difference in the binding between liver cytoplasm from embryos and adult rats is also considerable, yet much smaller than between the corresponding chromatins (Table 4). Our experiments have shown that the binding of hydrocortisone to cytoplasmic and chromatin receptors is rather specific so far as the unlabelled glucocorticoid competes for the binding sites with the labelled one, while the ability of other steroids to compete with 1,2$^3$H-hydrocortisone is much lower (Table 5). It is worth-while noting that the binding ability of rat liver chromatin considerably rises on the 3-4th day after birth though does not reach the adult level.

It has been reported recently by Tata (1970) that the administration of thyroid hormone to tadpoles induces metamorphosis only from 40 to 60 hours of development. It is noteworthy that the accumulation of this hormone in tadpole cells begins also only from this period.

Our experimental data, along with those in the literature, suggest that the formation of hormone receptors may transform definite groups of hormone-unsensitive cells (tissues, organs) into sensitive ones and ensure in this way the proper sequence of certain events underlying ontogenesis.

## REFERENCES

Cautrecasas, P. (1971). *Proc. natn. Acad. Sci. U.S.A.* **68**, 1264.
Fanestil, D. D. and Edelman, I. S. (1966). *Proc. natn. Acad. Sci. U.S.A.* **56**, 872.
Greenspan, F. S. and Hargadine, I. R. (1965). *J. Cell. Biol.* **26**, 177.
Karlson, P. (1963). *Angew. Chem.* **2**, 175.
Lavrinenko, I. A., Salganik, R. I. and Morozova, T. M. (1971). *Molek. biol.* **6**, 20.
Marushige, K. and Bonner, J. (1966). *J. molec. biol.* **15**, 160.
Morozova, T. M. and Salganik, R. I. (1969). *Molek. biol.* **3**, 745.
Noteboom, W. D. and Gorski, J. (1965). *Archs Biochem. Biophys.* **III**, 559.
Rodbell, M., Birnbauner, L. and Pohl, S. L. (1970). *J. biol. Chem.* **245**, 718.
Sereni, T., Kenney, F. T. and Kretchmer, N. (1959). *J. bch.* **234**, 609.
Swaneck, G. E., Chu, L. L. H. and Edelman, I. S. (1970). *J. biol. Chem.* **245**, 5382.
Tata, J. R. (1970). *Nature, Lond.* **227**, 686.
Toft, D., Shyamala, G. and Gorski, J. (1967). *Proc. natn. Acad. Sci. U.S.A.* **57**, 1740.
Yensen, E. V., Suzuki, T., Kawashima, I. S. and Stumpf, W. E. (1968). *Proc. natn. Acad. Sci. U.S.A.* **59**, 632.
Yeung, D., Stanley, D. S. and Oliver, I. T. (1967). *Biochem. J.* **105**, 1219.

FEBS Symposium, Volume 24, 1972, pp. 159-163

# Growth and Differentiation in Epidermal Cell Cultures from Embryonic Mouse Skin

N. E. FUSENIG, W. THON and S. M. AMER

*Deutsches Krebsforschungszentrum*
*Institut für Biochemie, Heidelberg, BDR*

Like other investigators interested in skin carcinogenesis, we have tried to establish a pure epidermal *in vitro* system for investigations on the influence of chemical carcinogens and co-carcinogens on proliferation and differentiation.

We used epidermal cells from term fetuses of $C_3H$-mice. After tryptic digestion of the skin, we eliminated the fibroblasts by sedimentations on three successive discontinuous Ficoll gradients. By this procedure we could separate cell clusters which consisted of epidermal cells, from single cells, mainly fibroblasts (Fusenig, 1971).

The cell clusters originated predominantly from embryonic hair follicles. They were incubated in Eagle's Medium in Falcon petri dishes, where they grew out to cell islands and finally formed a monolayer of polygonal cells (Fig. 1).

The proliferation rate of the cultures during the first days is expressed in Fig. 2 by the incorporation rates of $^{3}H$-thymidine, $^{3}H$-uridine, $^{14}C$-leucine and by the mitotic index. Following the initial peak after 30-36 h, the incorporation rates stay on the same level for at least six days, except for short-term increases following medium changes.

Maximum mitotic count in epidermal cells grown on glass cover-slips ranges from 0.4 to 0.6%. This is one-half to one-third the value found in hair follicles or stratum basale of the epidermis from term fetuses.

During the second day of cultivation a striking change in cellular morphology occurs with the appearance of filamentous structures in the cytoplasm. The frequently seen mitoses indicate that these alterations are not degenerative ones. On top of the original monolayer a second cell layer has formed which exhibits these cytoplasmic structures. This can clearly be observed in some areas of the cultures where the upper cell-sheet separates from the underlying cells, forming a transparent dome-shaped roof, through which the unchanged basal layer can be seen (Fig. 3).

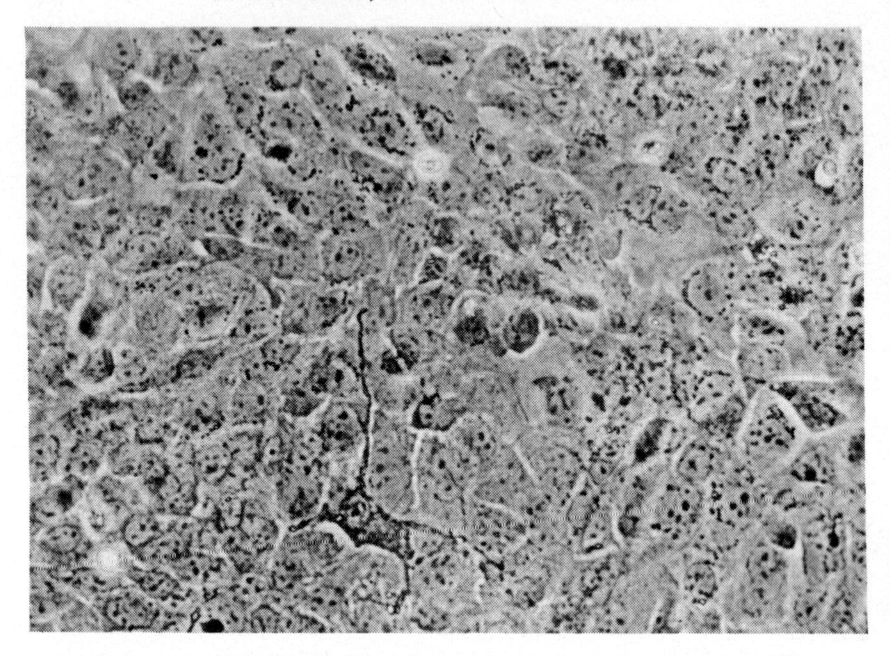

**Figure 1.** Epidermal cell-monolayer 24 h after plating; phase contrast × 150.

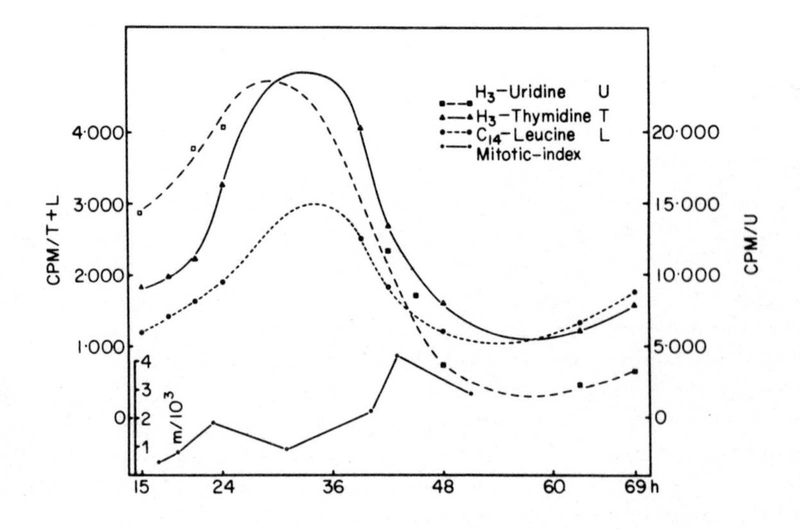


**Figure 2.** Epidermal cell cultures. Incorporation rates of $^{3}$H-thymidine, $^{3}$H-uridine and $^{14}$C-leucine and the mitotic index during the first three days after plating. (cpm/2,5 × $10^5$ cells; mitoses/$10^3$ cells).

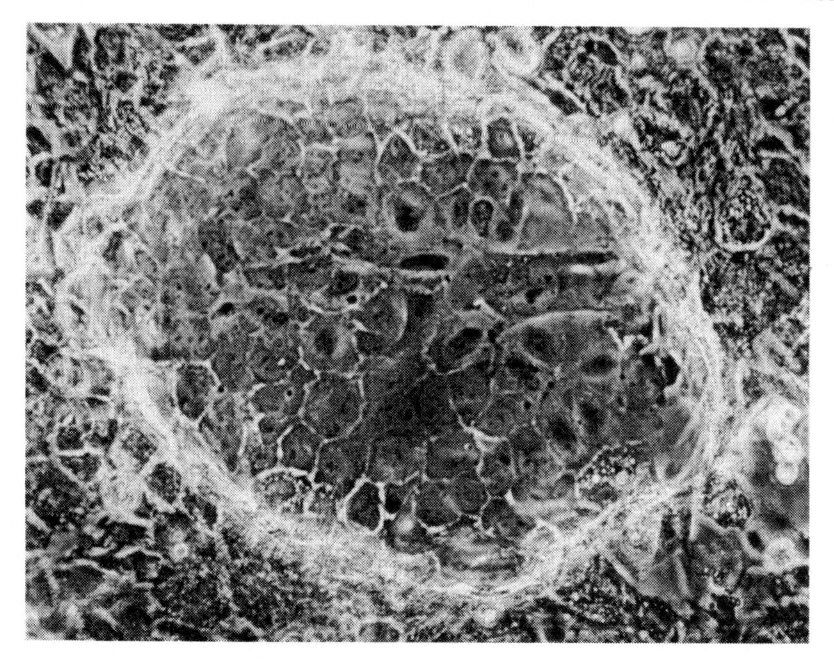

**Figure 3.** Epidermal cell culture, 48 h after plating. The upper fibrillar cell sheet has lifted from the basal layer, forming a transparent dome, and the unchanged basal layer is visible. Phase contrast x 150.

Three to four days after plating a new cell morphology appears on top of the cultures, exhibiting thickened, hexagonal membranes surrounding a transparent nucleus. This new cell type resembles hornified cells.

Six to ten day old cultures show in the uppermost cell layer large granulated cells with pyknotic nuclei (Fig. 4). Electronmicroscopic pictures (Voigt, personal communication) of vertical sections through 6 day old cultures clearly demonstrate that, starting from a monolayer, a culture has formed with at least 4 cell layers on top of the basal layer exhibiting the characteristic intercellular contacts by desmosomes. An epidermis-like structure had formed in these cultures in the absence of dermal elements.

The upper layers of these cultures can be separated by means of EDTA and trypsin leaving behind a proliferating basal layer, firmly attached to the plastic surface (Fig. 5). After staining with Papanicolaou and Rhodamin-B the isolated upper layers show the typical staining behaviour of hornified cells.

Further indications to the parakeratotic nature of these cells was brought about by their ability to show birefringence in the polarized light. Moreover, these *in vitro* formed horny sheets were not hydrolysed by 0.1 N Sodium hydroxyd or hydrochloric acid.

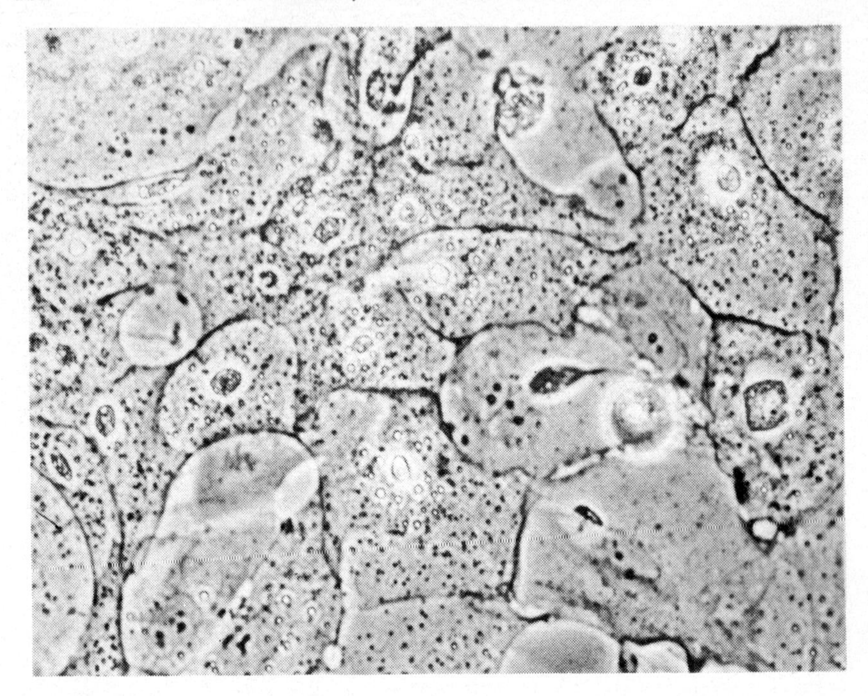

**Figure 4.** Epidermal cell culture, 8 days after plating. The uppermost layer of enlarged, granulated cells with pyknotic nuclei is focused. Phase contrast x 250.

Besides the morphological and histochemical features of the parakeratotic-like hornification of epidermal cells in culture, there are further indications that differentiation and specific epidermal functions are maintained in these cultures.

At different times after plating we could measure in the epidermal cultures, activities of histidase ($4 \times 10^{-3}$ $\mu$M/min/mg Protein)–an enzyme specific for epidermis (Barnhisel *et al.*, 1970)–comparable to those found in human foreskin epidermis.

By means of a specific anti-epidermal-antiserum, Dr. P. Worst could prove, with the mixed hemadsorption technique, that the epidermal cells in culture exhibit epidermis-specific histocompatibility antigens (Worst and Fusenig, in preparation).

Bromodeoxyuridine (BUDR) in concentrations which do not inhibit DNA synthesis ($10^{-8}$ M), blocks the development of parakeratotic upper layers. The BUDR-treated cultures could be maintained for up to four weeks without the appearance of hornification.

Vitamin-A-acid (VAS) in concentrations up to 0.5 $\mu$g/ml leads to a 30% stimulation of DNA-synthesis in epidermal cells *in vitro*. In the same

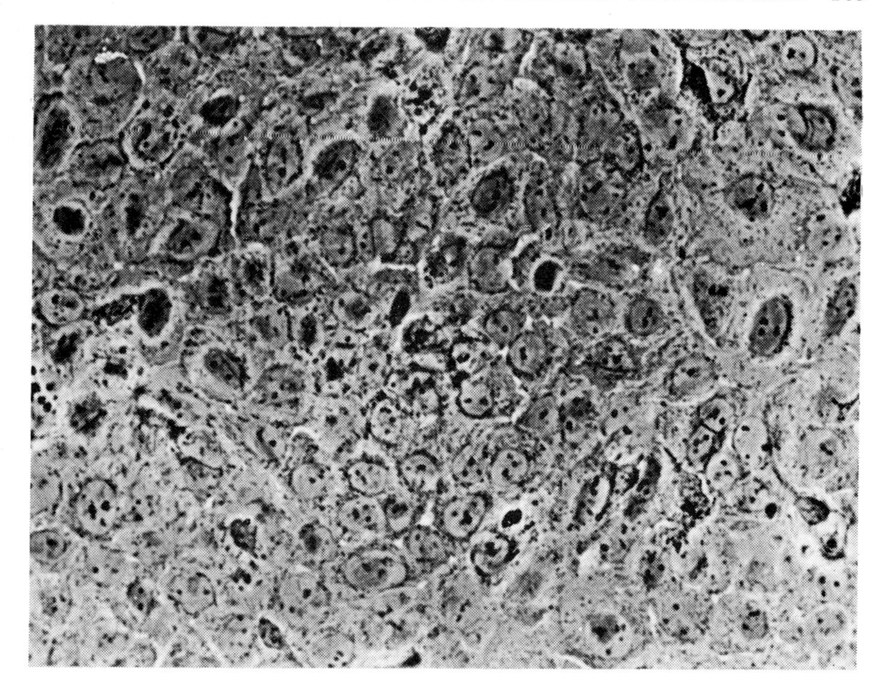

**Figure 5.** Epidermal cell culture 6 days after plating, showing the basal layer which is left after separation of the parakeratotic upper layers. Phase-contrast × 150.

concentrations VAS inhibits differentiation. These results are comparable to the findings of Christophers and Braun-Falco (1968) derived from ear epidermis of guinea pigs.

Commercially available epidermal chalone from pig skin which inhibits specifically epidermal DNA-synthesis and mitoses *in vivo*, reduces the thymidine incorporation to 50% after 6 h incubation.

We consider these epidermal cell cultures in their proliferative and differentiative capacities, which are maintained *in vitro* without dermal influence, a useful model system for biochemical investigations on differentiation and on the disturbance of differentiation by carcinogenic and co-carcinogenic substances.

## REFERENCES

Barnhisel, M. L., Priest, R. L. and Priest, J. H. (1970). *J. cell comp. Physiol.* **76**, 7.

Christophers, E. and Braun-Falco, O. (1968). *Arch. klin. exp. Derm.* **232**, 427.

Fusenig, N. E. (1971). *Naturwissenschaften.* **58**, 421.

FEBS Symposium, Volume 24, 1972, pp. 165-167

# Membrane Events and Liver Regeneration

MARGERY G. ORD and L. A. STOCKEN

*Department of Biochemistry, University of Oxford, Oxford, England*

An early event in liver regeneration is the appearance of ornithine decarboxylase (E.C.4.1.1.17) (Jänne and Raina, 1968; Russell and Snyder, 1968). The enzyme catalyses the formation of putrescine which is rate-limiting for the provision of spermine needed for the assembling of increased numbers of polysomes after partial hepatectomy. The rapidity of the enzyme induction suggested that it might be promoted, *inter alia*, by cyclic AMP; our un-

**Table 1.** Effect of phenoxybenzamine on the induction of ornithine decarboxylase and DNA synthesis in regenerating rat liver.

| | Ornithine decarboxylase | DNA content |
|---|---|---|
| *Controls* | | |
| | 4 h | 22 h |
| Sham-operated | 1.4 ± 0.7 | 1.1 ± 0.19 |
| Partially hepatectomized | 23.5 ± 13.9 | 1.5 ± 0.26 |
| Adrenalectomized, partially hepatectomized | 6.2 ± 4.5 | 1.6 ± 0.1 |
| *Phenoxybenzamine treated, partially hepatectomized rats* 5 mg/kg | | |
| −0.25 h | 2.5 ± 1.7 | 1.8 ± 0.32 |
| +5 h | – | 1.3 ± 0.01 |
| +21 h | – | 1.8 ± 0.48 |
| *Phenoxybenzamine treated, partially hepatectomized rats* 10 mg/kg | | |
| − 0.25 h | 2.2 ± 0.6 | 1.1 ± 0.39 |
| +4 h | – | 1.1 ± 0.16 |
| +21 h | – | 1.4 ± 0.32 |

Ornithine decarboxylase levels were measured using [1-$^{14}$C]ornithine (Russell and Snyder, 1968) and are given as nmoles ornithine decarboxylated/g soluble protein/0.5 h at 37°C. DNA content was measured 22 h after operation and is given as mg DNA/g liver wet weight. The phenoxybenzamine was administered intraperitoneally at the times indicated relative to partial hepatectomy.

published observation showed that levels of cyclic AMP might be increased in livers of partially hepatectomized rats 1.5 h after operation.

The pattern of ornithine decarboxylase induction in our rats was similar to that reported by Russell and Snyder (1968) (Table 1). Adrenalectomy, which would decrease the level of glucocorticoids and prevent activation of adenyl cyclase through $\beta$ receptors (Robison, *et al.*, 1971), reduced the amount of decarboxylase detectable at 4 h (contrast Russell and Snyder, 1969) but did not delay DNA synthesis measured at 22 h in the midpoint of the S period. Propranolol, which blocks $\beta$ sites, given immediately before operation had no effect on the induction nor on DNA synthesis, but chlorpromazine and phenoxybenzamine, which are $\alpha$ blockers, inhibited the induction of ornithine decarboxylase and subsequent DNA synthesis. 1 mg/kg phenoxybenzamine reduced the amount of enzyme found at 4 h by 50% and complete inhibition of the induction was produced by doses of 5 mg/kg and above (Table 1).

When 5 mg/kg phenoxybenzamine were given immediately before operation the effects on decarboxylase induction were transient since amounts of enzyme at 22 h were in the same range as those in untreated partially hepatectomized rats. Preliminary observations showed that if the drug was given at 5 h, enzyme was present at 22 h, in amounts rather lower than in untreated rats.

Phenoxybenzamine also affected DNA synthesis after partial hepatectomy. At 10 mg/kg the drug was inhibitory when given at –0.25, 4 or 21 h after operation. When examined after administration at –0.25 h, the inhibition was temporary since by 26 h the DNA content had risen and, at 2 and 3 weeks, treated animals had liver and body weights undistinguishable from untreated, partially hepatectomized rats. At 5 mg/kg DNA synthesis was only reduced when the drug was given 5 h after operation.

Although the site of action of phenoxybenzamine is uncertain the data suggest that $\alpha$ receptor sites are involved in early events in liver regeneration leading to the induction of ornithine decarboxylase. The period during which the enzyme is present after operation implies that the induction is maintained over a number of hours (Panko and Kenney, 1971). It is not possible from these results to dissociate events leading to increased polysome formation from subsequent DNA synthesis but the effects of 5 mg/kg phenoxybenzamine 5 h after operation indicate that plasma membrane sites may still be involved at this time.

## ACKNOWLEDGEMENTS

We are grateful to the Cancer Research Campaign for financial assistance and to Messrs. Smith, Kline and French for the phenoxybenzamine and for information regarding its usage.

## REFERENCES

Jänne, J. and Raina, A. (1968). *Acta chem. scand.* **22**, 1349.

Panko, W. B. and Kenney, F. T. (1971). *Biochem. biophys. Res. Commun.* **43**, 346.

Robison, G. A., Butcher, R. W. and Sutherland, E. W. (1970). *In* Fundamental Concepts of Drug-Receptor Interactions p. 59 (J. F. Danielli, J. F. Moran, and D. J. Triggle, eds), Academic Press, New York and London.

Russell, D. H. and Snyder, S. H. (1968). *Proc. natn. Acad. Sci. U.S.A.* **60**, 1420.

Russell, D. H. and Snyder, S. H. (1969). *Endocrinology* **84**, 223.

FEBS Symposium, Volume 24, 1972, pp. 169-173

# The Pathways of Carbohydrate Oxidation in Growing and Dividing Cells of *Acer pseudoplatanus L.* (Sycamore) Grown in Batch Culture

M. W. FOWLER*

*Botanical Laboratories, School of Biology, University of Leicester, Leicester, U.K.*

The rapid development of plant cell culture techniques during the last few years has brought to the fore a system which shows great potential as a tool in the investigation of biochemical differentiation in plant cells. Using such a system we have carried out an investigation into the role played by the EMP (Embden-Meyerhof-Parnas) pathway and the pentose phosphate pathway in the metabolism of growing and dividing sycamore cells.

Sycamore cells, in common with cultured cells from a number of other plant species, exhibit certain characteristic stages in their growth when maintained in batch culture (Wilson *et al.*, 1971) (Fig. 1). Although showing no obvious signs of differentiation into distinct cell types (e.g. xylem or phloem cells) during growth on the particular medium which we use (Fowler, 1971), the cells do in fact undergo marked changes in their fine structure and form (Sutton-Jones and Street, 1968; Davey and Street, 1971). The occurrence of such changes in fine structure, together with the ease of maintenance of sycamore cell cultures prompted their use in our investigations.

If we view differentiation as an ordered, sequential expression of fractions of the cell genome, then the changing catalytic capacities of metabolic pathways as seen in the change in enzyme activities constitute an important aspect of the differentiation process. In consequence attention was paid to the activities of several enzymes of the two pathways of carbohydrate oxidation during growth of the cells. Particular attention was given to phosphofructokinase in the EMP pathway, and to glucose-6-phosphate dehydrogenase in the pentose phosphate pathway. These two enzymes are unique to their respective pathways and appear to be major sites of regulation within the reaction

* Present address: Department of Biochemistry, University of Sheffield, Sheffield, 10. U.K.

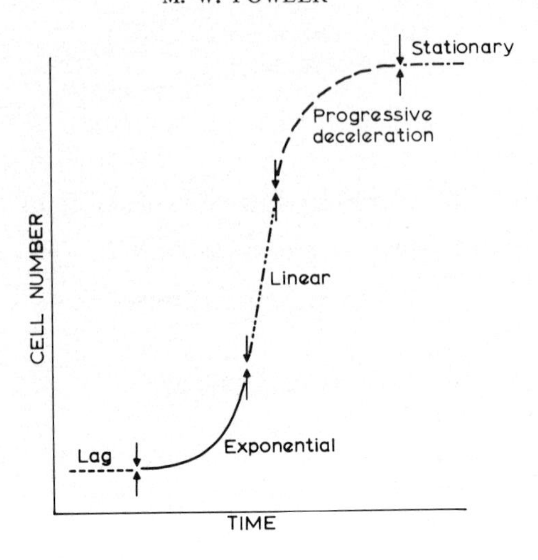


**Figure 1.** Model curve relating cell number per unit volume of culture to time in a batch grown plant cell suspension culture. Growth phases labelled. (From Wilson *et al.*, 1971).

sequences (Fowler, 1969). The changes in activity of the different enzymes during the growth of the cells is shown in Fig. 2 (A, B, C).

During the first few days of growth after culture initiation the activity of glucose-6-phosphate dehydrogenase showed a much greater increase, and for a longer period of time, than the activity of phosphofructokinase. Following the initiation of cell division, at about day 4, this pattern of enzymic activity changed markedly. The activity of glucose-6-phosphate dehydrogenase decreased significantly, reaching a similar level to that of phosphofructokinase, which had also decreased slightly.

To assess whether or not the changes in the relative catalytic capacities of the two pathways were accompanied by changes in their relative utilization, a series of isotopic experiments were carried out using [1-$^{14}$C] and [6-$^{14}$C] glucose. The $CO_2$ released by the cells during incubation in the labelled substrate was collected and assayed for radioactivity. The data were then analysed by the Bloom and Stetten $^{C6}/C_1$ ratio method (Bloom and Stetten, 1953), and a summary of the data is presented in Table 1. A $^{C6}/C_1$ ratio of unity is taken to indicate that carbohydrate oxidation is mainly through EMP pathway. Ratios lower than unity are taken to indicate that both pathways are making a significant contribution to carbohydrate oxidation. It must of course, be noted that the very complexity of carbohydrate oxidation precludes anything but a very general indication of the true position being obtained by this method.

**Table 1.** The release of $^{14}C$ as $CO_2$ from $(1\text{-}^{14}C)$- and $(6\text{-}^{14}C)$-glucose supplied to harvested cells from suspension cultures of *Acer pseudoplatanus* at different stages of growth. Cells incubated in isotope for 45 min.

| Days after culture inoculation | Position of label in glucose | Percentage of added $^{14}C$ released as $CO_2$ | C6/C1 |
|---|---|---|---|
| 0 | C-1 | 6 | 0.50 |
| | C-6 | 3 | |
| 1 | C-1 | 7 | 0.35 |
| | C-6 | 2 | |
| 2 | C-1 | 13 | 0.31 |
| | C-6 | 4 | |
| 4 | C-1 | 19 | 0.42 |
| | C-6 | 8 | |
| 7 | C-1 | 10 | 0.70 |
| | C-6 | 7 | |
| 10 | C-1 | 20 | 0.75 |
| | C-6 | 15 | |
| 14 | C-1 | 14 | 0.85 |
| | C-6 | 12 | |
| 20 | C-1 | 17 | 0.88 |
| | C-6 | 15 | |

Study of the data in Table 1 suggests that during the first few days of growth, before cell division commences, carbohydrate oxidation was through both the EMP pathway and the pentose phosphate pathway. Following day 4 of incubation, and the initiation of cell division, carbohydrate oxidation gradually became predominantly via the EMP pathway.

From the above data it is suggested that the changes in the relative catalytic capacity of the two pathways are in turn accompanied by associated changes in the relative carbon fluxes of the two pathways. Under the particular growth conditions prevailing, the operation of the pathways therefore appears to be under what Kornberg (1966) has called "coarse control".

The general interpretation placed upon the data in relation to growth of the cells is as follows. In the short period of 3-4 days following culture initiation the metabolic activities of the culture are geared to biosynthesis (Fowler, 1971; Wilson, 1971). The high relative activity of the pentose phosphate pathway in relation to the EMP pathway is thought to be related to the provision of NADPH by the former pathway in support of this biosynthesis. At this time the cells in the culture are densely cytoplasmic with

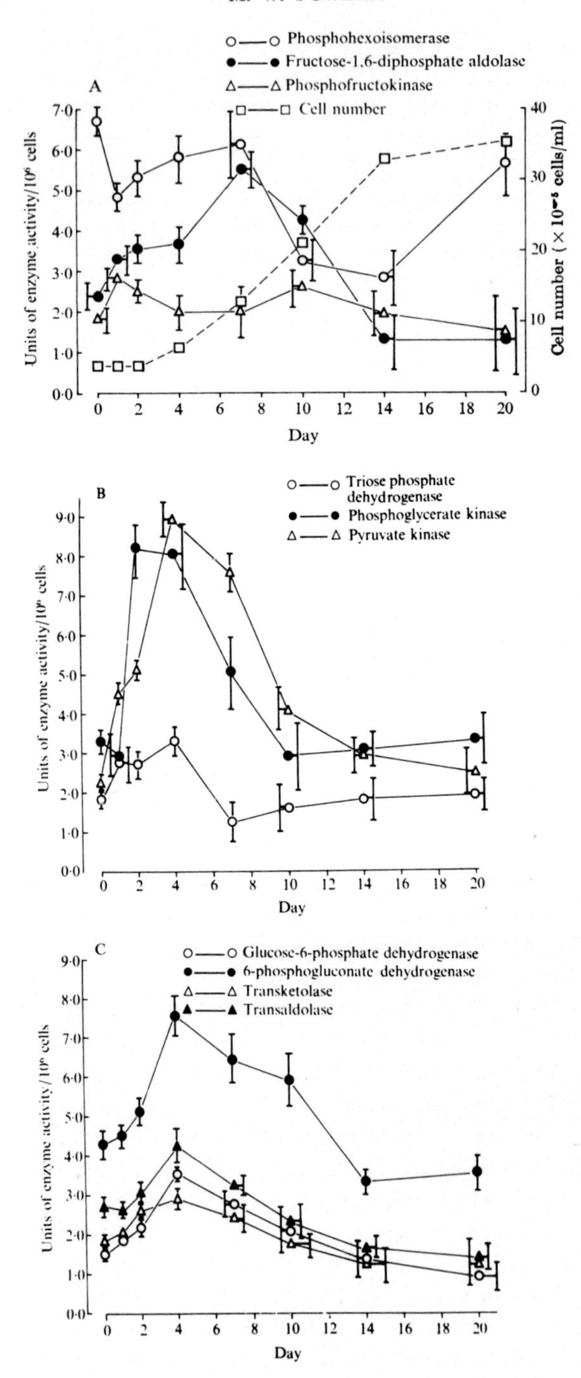
A
Phosphohexoisomerase
Fructose-1,6-diphosphate aldolase
Phosphofructokinase
Cell number
Units of enzyme activity/10⁶ cells
Cell number (× 10⁻⁵ cells/ml)
Day
B
Triose phosphate dehydrogenase
Phosphoglycerate kinase
Pyruvate kinase
Units of enzyme activity/10⁶ cells
Day
C
Glucose-6-phosphate dehydrogenase
6-phosphogluconate dehydrogenase
Transketolase
Transaldolase
Units of enzyme activity/10⁶ cells
Day

an increasing number of ribosomes and mitochondria. The cultures show little evidence of cell division. Studies by Dr. Wilson in our laboratory also suggest that at this time the citric acid cycle operates in concert with the pentose phosphate pathway, providing some of the carbon skeletons necessary for the biosynthetic activity. After the initiation of cell division at about day 4, the cells become more vacuolate and less dense, and there is a concomitant increase in lipoidal material amd membrane area. The balance of metabolism appears to move away from biosynthesis, towards polymerization. Certain evidence of this comes from the relative decrease in the activity of the pentose phosphate pathway compared with the EMP pathway, and readjustment of the pathways of mitochondrial oxidative phosphorylation towards ATP formation.

One final point here which is of interest in relation to gene expression and enzyme activity within the cultures. Study of the data for the pentose phosphate pathway enzymes shows that the relative activities of the enzymes assayed remained similar throughout the growth period. Pette *et al.*, (1962) suggested that enzymes acting in this way may form genetically regulated "constant proportion groups". The occurrence of this phenomena does not however appear to be universal (Fowler and ap Rees, 1970). At present we are trying to obtain further information on this point.

## ACKNOWLEDGEMENTS

The author thanks Professor H. E. Street for advice and encouragement. The work was carried out during the tenure of an I.C.I. Research Fellowship.

## REFERENCES

Bloom, G. and Stetten, D. (1953). *J. Am. Chem. Soc.* 75, 5446.
Davey, M. R. and Street, H. E. (1971). *J. exp. Bot.* 22, 90.
Fowler, M. W. (1969). Ph.D. Thesis. University of Cambridge.
Fowler, M. W. (1971). *J. exp. Bot.* **22**, 715.
Fowler, M. W. and ap Rees, T. (1970). *Biochim. biophys. Acta.* **201**, 33.
Kornberg, H. L. (1966). *Biochem. J.* **99**, 1.
Pette, D., Luh, W. and Bucher, T. L. (1962). *Biochem. biophys. Res. Commun.* 7, 414.
Sutton-Jones, B. and Street, H. E. (1968). *J. exp. Bot.* **19**, 114.
Wilson, S. B. (1971). *J. exp. Bot.* **22**, 725.
Wilson, S. B., King, P. J. and Street, H. E. (1971). *J. exp. Bot.* **22**, 177.

---

**Figure 2.** A.B.C. Growth of Sycamore cells in batch suspension culture, (cell number/ml culture), and activities of enzymes of the EMP pathway and pentose phosphate pathway in extracts of sycamore cells grown in suspension culture.

A unit of enzyme activity is defined as the utilization of 1 *n* mole of substrate/min at 25°C. The lengths of the vertical lines represent twice the standard errors of the mean values plotted in the mean as mid-point. The activities of phosphohexoisomerase and phosphoglycerate kinase presented are one-half of the measured activities of the enzymes.

# Cellular Differentiation and Secondary Metabolism of Microorganisms

M. LUCKNER and L. NOVER

*Sektion Pharmazie of the University of Halle and Institut für Biochemie der Pflanzen of the DAW, Halle (Saale), DDR*

Secondary metabolites, e.g. antibiotics, alkaloids, polyketides etc., are formed by many microorganisms and higher plants. They usually are synthesized only during certain developmental stages of these organisms as a consequence of cell differentiation processes. The simple and sensitive methods for the measurement of biosynthesis and breakdown of secondary metabolites which are available provide a good basis of the investigation of regulatory processes. Furthermore, the manifestation of secondary metabolism often coincides with steps of morphological differentiation, thus offering the possibility of investigating relations between "chemical" and "morphological" differentiation.

As an example in this respect the secondary metabolism and the morphogenesis of *Penicillium cyclopium* and *P. viridicatum* were studied. After a period of intensive growth (trophophase) these moulds undergo a differen tiation process which is characterized by the formation of conidiospores and of alkaloides of the benzodiazepine-quinoline type (idiophase).

FEBS Symposium, Volume 24, 1972, pp. 177-179

# Cellular Reprogramming and Cellular Differentiation

R. TSANEV

*Biochemical Research Laboratory*
*Bulgarian Academy of Sciences*
*Sofia, Bulgaria*

In this Symposium many aspects of the emergence of tissue specific proteins during development of eukaryotes have been discussed. However, from the point of view of the molecular mechanisms controlling this process, there is an important question which has not been raised at all. The problem is whether all phenomena where the synthesis of new protein species is switched on without changes in DNA should be mixed under the same label of cellular differentiation, as has been implicitly done during this Symposium or stated in the literature by some authors (Jacob and Monod, 1963).

I think it is very important to distinguish between at least two groups of biological phenomena involved in the synthesis of new protein species. Both groups can be defined as *cellular reprogramming*—a process when DNA remains unchanged but the synthetic pattern of the cell is altered. A detailed analysis of different cases of cellular reprogramming clearly shows that two groups can be distinguished which differ in their biological characteristics (Tsanev and Sendov, 1971a, b). The first group can be considered as changes in the *functional state* of the cell while the second involves changes in the *cellular type.*

The first group includes processes of reversible enzyme induction in prokaryotic and eukaryotic cells, injury-induced cellular proliferation, environment-induced reversible transformations in some protozoa as is the interesting case of *Naegleria* (Willmer, 1963). The second group is represented by the emergence, during embryonic development, of different cellular types determined to synthesize different proteins.

As can be seen from Table 1 the biological characteristics of these two groups are not only different but quite opposite. Obviously they cannot be controlled by the same molecular mechanisms.

It would be easy to explain the characteristics of the first group on the basis of processes of *repression-derepression* based on a reversible, concentration dependent binding of repressors to DNA. However, the biological

**Table 1.** Cellular reprogramming*

1. Altered protein pattern
2. No changes in DNA

| I. Changes in the functional state of the cell | II. Changes in the cellulary type |
|---|---|
| 1. Reversible | 1. Irreversible under normal anatomical and physiological conditions |
| 2. Monophasic | 2. Multiphasic |
| 3. Short latent period | 3. Long latent period |
| 4. Inducer permanently needed | 4. Inducer temporarily needed |
| 5. Mitosis independent | 5. Mitosis dependent |

* For details see Tsanev and Sendov (1971a).

characteristics of the second group require a principally different mechanism which we tentatively called *blocking-deblocking* (Tsanev and Sendov, 1971a) and which involves an irreversible binding of molecules (most probably chromatin proteins) to DNA. Analysing such a model we have shown that the final stage of cellular differentiation can only be reached if we suppose a chain of events where both processes of repression-derepression and of blocking-deblocking are successively involved (Tsanev and Sendov, 1971a). The need for at least two repressor systems in cellular differentiation has also been felt by other authors (Holoubek *et al.*, 1966; Paul, 1967; Stellwagen and Colle, 1969).

In a model we have recently proposed (Tsanev and Sendov, 1971a), DNA of eukaryotic cells is supposed to be blocked by histones which form specific arrangements at different sites of the genome. In this way different genetic units can be specified and becomes recognizable by another group of proteins—the nonhistone proteins of the chromatin. These proteins are bound irreversibly, but specifically, to different combinations of histones and thus deblock the corresponding sites of the genome. According to this model, a cell with a modified profile of deblocking proteins in its chromatin can only be obtained through the combined effect of a mitotic division and of some factor (or inducer) which can repress the synthesis of some deblocking proteins during the mitotic cycle.

In this way the irreversible binding of some deblocking proteins to the chromatin can explain the requirement for mitotic divisions to change the cellular type (Holtzer *et al.*, 1973), while the induction of different synthetic programmes can be easily explained by processes of repression-depression where mitotic divisions are not necessary (Moscona, 1972). The clear distinction between these two groups of phenomena will be helpful both in explaining some disagreements and the molecular basis of cellular differentiation.

## REFERENCES

Holoubek, V., Fanshier, L., Crocker, T. T. and Hnilica, L. S. (1966). *Life Sci.* **5**, 1691.
Holtzer, H., Weintraub, H. and Biehle, J. (1973). This Symposium, p. 41.
Jacob, F. and Monod, J. (1963). *In* Cytodifferentiation and Macromolecular Synthesis (M. Locke, ed.), p. 30, Academic Press, New York and London.
Moscona, A. A. (1972). this Symposium.
Paul, J. (1967). *In* Cell Differentiation (A. V. S. DeReuk and J. Knight, eds), p. 196. Ciba Found. Symp., Churchill, London.
Stellwagen, R. H. and Colle, R. D. (1969). *A. Rev. Biochem.* **38**, 951.
Tsanev, R. and Sendov, Bl. (1971a). *J. theoret. Biol.* **30**, 337.
Tsanev, R. and Sendov, Bl. (1971b). *Z. Krebsforsch.* **76**, 299.
Willmer, E. N. (1963). *Symp. Soc. exp. Biol.* **17**, 215.

# Author Index

Numbers followed by an asterisk are those pages on which references are listed.

D



E



F



G

**R**



**S**



**T**

**V**



**W**



**Y**



**Z**

# Subject Index

D



E



F



G

**S**



**T**



**U**



**V**